Experimental Investigation of Calcium Looping CO₂ Capture for Application in Cement Plants

Matthias Hornberger

Experimental Investigation of Calcium Looping CO$_2$ Capture for Application in Cement Plants

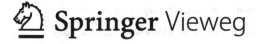

Springer Vieweg

Matthias Hornberger
Institute of Combustion and Power Plant
Technology
University of Stuttgart
Stuttgart, Germany

Zugl.: Dissertation Universität Stuttgart, 2022

D93

ISBN 978-3-658-39247-5   ISBN 978-3-658-39248-2   (eBook)
https://doi.org/10.1007/978-3-658-39248-2

Responsible Editor: Stefanie Probst
This Springer Vieweg imprint is published by the registered company Springer Fachmedien
Wiesbaden GmbH, part of Springer Nature.
The registered company address is: Abraham-Lincoln-Str. 46, 65189 Wiesbaden, Germany

Acknowledgement

The work presented in this thesis has been conducted within my employment at the Institute of Combustion and Power Plant Technology of the University of Stuttgart between 2015 and 2021. Most of the presented work is related to two European research projects *CEMCAP* and *CLEANKER* funded by the Horizon 2020 research and innovation programme. I like to express my gratitude to Prof. Scheffknecht for the opportunity to conduct my research at the Institute of Combustion and Power Plant Technology (IFK) and its unique infrastructure. Moreover, I want to thank Prof. Scheffknecht for his scientific supervision, guidance and support. My gratitude is extended to Prof. Dr.-Ing. Alexandros Charitos who agreed to act as co-examiner.

I am grateful for the support of my colleagues at the Institute of Combustion and Power Plant Technology in various aspects. I deeply thank Heiko Holz for his vast technical and moral support. I further want to thank Reinhold Spörl and Gebhard Waizmann for the inspiring discussions also exceeding the topic of my research work. I appreciate the professional exchange with Joseba Moreno and his willingness to help. Also the valuable help of Dieter Straub in all matters is greatly appreciated. A special thanks goes to the department of Decentralised Energy Conversion for the extensive support during the experimental campaigns: Theodor Beisheim, Andreas Gredinger, Gerrit Hofbauer, Henning Luhmann, Joseba Moreno, Marcel Beirow, Daniel Schweitzer, Max Schmid, Tim Seitz, Max Weidmann and Heiko Dieter.

Additional thanks go to Wolfgang Ross and his team at the 'Laboratory for Fuels, Ashes and Slag' at the IFK for supporting me with analyses and their experimental equipment and to IFK's work shop team (Ralf Nollert and Thomas Froschmeier) for construction and modification of experimental equipment. I appreciate assistance of Antje Radszuweit, Beate Koch, Renate Klein, Claus Nagel and Marja Steinlechner regarding all kind of administrative and IT issues.

Lastly, I want to thank my family and friends for their support. I am blessed to have friends like you in my life. I am especially grateful for my mother Martina Hartmann and her unconditional support and encouragement to pursue my goals in life. Finally, I want to dearly thank my partner Sophia Bruttel, who has been an abundant source of high spirits and a caring supporter.

Stuttgart, 2021
Matthias Hornberger

Abstract

This thesis assesses the application of the calcium looping technology for CO_2 capture from cement plants. The cement industry contributes significantly to the anthropogenic CO_2 emissions. Due to process inherent CO_2 emissions, the application of CCS technologies is inevitable to fully decarbonise the cement sector and mitigate climate change. Within this thesis, various integration options of the calcium looping technology into the cement clinker manufacturing process have been developed addressing different boundary conditions of the cement plants. The more mature options using fluidised bed reactors have been extensively studied at semi industrial scale for operation conditions anticipated for the respective integration option. Furthermore, a novel concept using entrained flow reactors has been assessed by investigating the sorbent properties of various raw meals in such a system. Besides, a comprehensive study regarding the suitability of various potential sorbents (i.e. limestone, raw meal and raw meal components) have been conducted using thermogravimetric analysis.

The comprehensive sorbent screening showed that raw meal based sorbents suffer a severe deactivation during the first calcination that can be attributed to belite formation. The severity of the initial sorbent deactivation increased slightly with increasing SiO_2 content of the sorbent. Short calcination times enhanced the sorbent activity as the formation of belite is limited. At oxy-fuel conditions, all raw meal based sorbents show very similar cyclic CO_2 carrying capacities that can be described by a single deactivation model. For a calcination time of 10 min, the raw meals' CO_2 carrying capacity can be described using the deactivation model of Ortiz et al. with an initial CO_2 carrying capacity of $0.30 \, mol \, mol^{-1}$, a deactivation constant of 0.44 and a residual activity of $0.075 \, mol \, mol^{-1}$. Whereas, a calcination time of 1 min yields an initial CO_2 carrying capacity of $0.39 \, mol \, mol^{-1}$, a deactivation constant of 0.28 and a residual activity of $0.091 \, mol \, mol^{-1}$. Assessing various raw meal qualities at entrained flow conditions showed that the calcination degree increased with increasing calcination temperature and increasing residence time. The recarbonation showed no dependence of the calcination temperature in the investigated temperature range (i.e.

900 °C and 920 °C) but decreased slightly with increased residence time. Overall, the re-carbonation performances of entrained flow calcined raw meals are in agreement with the recarbonation behaviour of limestone based sorbents. The CO_2 carrying capacity increased with increasing carbonation temperature, whereas the CO_2 concentration did not affect the sorbent conversion but its carbonation reaction rate. For CO_2 concentrations below $0.2 \, m^3 \, m^{-3}$, a reaction order of 1 regarding the driving force was found. The investigation of the fluidised bed calcium looping CO_2 capture options yielded CO_2 capture efficiencies up to 98 % due to the significantly increased sorbent activity. For high integration levels CO_2 capture in the carbonator was limited by the achievable equilibrium CO_2 concentration. At lower integration levels CO_2 capture was limited by the circulated amount of active CaO. The CO_2 capture could be adjusted by altering the circulation rate. Overall, the CO_2 capture efficiency in the fluidised bed carbonator can be described by the active space time model developed for power plant application. Based on the experiments a CO_2 purity of $0.95 \, m^3 \, m^{-3}$ can be anticipated for the oxy-fuel calciner's flue gas before compression and purification (i.e. before the CPU) with excess oxygen being the main side component. Due to the large quantity of calcium in the calciner, sulphurous components were completely captured from the flue gas, whereas NO_x formation increased significantly. NO_x concentration increased with excess oxygen and ranged from $500 \cdot 10^{-6} \, m^3 \, m^{-3}$ to $750 \cdot 10^{-6} \, m^3 \, m^{-3}$ requiring a dedicated NO_x removal step during compression and purification.

Kurzfassung

Diese Arbeit untersucht die Implementierung des Calcium-Looping-CO_2-Abscheideverfahrens in den Zementklinkerherstellungsprozess. Die Zementindustrie trägt signifikant zu den anthropogenen CO_2-Emissionen bei. Da der Großteil der CO_2-Emissionen in der Zementindustrie prozessinhärent ist, müssen CCS-Technologien eingesetzt werden, um die Zementklinkerproduktion vollständig zu dekarbonisieren. Unter Berücksichtigung der vorherrschenden Rahmenbedingungen des Zementwerks wurden im Rahmen der vorliegenden Arbeit unterschiedliche Integrationsmöglichkeiten zwischen dem Calcium-Looping-Prozess und der Klinkerproduktion erarbeitet. Eine ausgereiftere Variante, bei der Wirbelschichtreaktoren eingesetzt werden, wurde im halbtechnischen Maßstab umfassend für Betriebsbedingungen diverser Integrationsgrade untersucht. Eine neuartige Variante, bei der Flugstromreaktoren zum Einsatz kommen, wurde im Rahmen einer Untersuchung zum Verhalten unterschiedlicher europäischer Rohmehlqualitäten in einem solchen System untersucht. Darüber hinaus wurde eine ausführliche thermogravimetrische Studie zur Eignung diverser Ausgangsmaterialien (Kalksteine, Rohmehle und Rohmehlkomponenten) als Calcium-Looping-Sorbens durchgeführt.

Die Studie zur Eignung potentieller Sorbenzien ergab, dass rohmehlbasierte Sorbenzien einer starken Deaktivierung während der ersten Kalzinierung ausgesetzt sind, welche auf die Bildung von Belite zurückzuführen ist. Das Ausmaß der Deaktivierung steigt mit zunehmenden SiO_2-Gehalt an. Durch kurze Kalzinierungszeiten kann die Sorbensaktivität erhöht werden, da das Ausmaß der Belitbildung reduziert wird. Unter oxy-fuel-Bedingungen zeigten alle untersuchten Sorbenzien übereinstimmend eine vergleichbare CO_2-Aufnahmekapazität. Für eine Kalzinierungszeit von 10 min kann die zyklische CO_2-Aufnahmefähigkeit nach dem Model von Ortiz et al. mit einer Anfangsaktivität von $0{,}30\,\mathrm{mol\,mol^{-1}}$, einer Deaktivierungskonsante von 0,44 und einer Residualaktivität von $0{,}075\,\mathrm{mol\,mol^{-1}}$ beschrieben werden. Eine reduzierte Kalzinierungszeit von 1 min ergab eine Anfangsaktivität von $0{,}39\,\mathrm{mol\,mol^{-1}}$, eine Deaktivierungskonsante von 0,28 und eine Residualaktivität von $0{,}091\,\mathrm{mol\,mol^{-1}}$.

Die Untersuchung der verschiedenen Rohmehlqualitäten unter Flugstrombedingungen ergab, dass der Kalzinierungsgrad mit zunehmender Kalzinierungstemperatur und zunehmender Verweilzeit ansteigt. Im untersuchten Temperaturbereich (900 °C und 920 °C) wurde keine Abängigkeit des Rekarbonatisierungsgrads von der Kalzinierungstemperatur festgestellt. Mit zunehmender Verweilzeit nahm die CO_2-Aufnahmefähigkeit jedoch leicht ab. Insgesamt zeigten flugstromkalzinierte Rohmehle das gleiche Rekarbonatisierungsverhalten wie kalksteinbasierte Sorbenzien. Mit zunehmender Karbonatisierungstemperatur nahm die CO_2-Aufnahmefähigkeit zu. Eine Erhöhung der CO_2-Konzentration beeinflusste den Karbonatisierungsumsatz nicht, führte jedoch zu einer Erhöhung der Karbonatisierungsgeschwindigkeit. Für CO_2-Konzentrationen unter 0,2 m^3 m^{-3} wurde eine Reaktionsordnung ersten Grades bezüglich des treibenden Konzentrationsgefälles festgestellt.

Die Untersuchungen zur Wirbelschichtvariante des Calcium-Looping-Prozesses ergaben CO_2-Abscheideeffizienzen bis zu 98 % dank einer stark erhöhten Sorbenzaktivität. Für hohe Integrationsfälle wurde die CO_2-Abscheidung durch die erreichbare Gleichgewichtskonzentration des CO_2 beschränkt. Im Falle niedrigerer Integrationsgrade war die CO_2-Abscheidung durch den eintretenden Strom an aktiven CaO limitiert und konnte durch eine Änderung der Zirkulationsrate angepasst werden. Die CO_2-Abscheidung im Karboantor konnte mit guter Übereinstimmung mit dem Acitve Space Time Model, das ursprünglich für die Kraftwerksapplikation entwickelt wurde, beschrieben werden. Auf Basis der Ergebnisse kann geschlussfolgert werden, dass die CO_2 Reinheit des Abgases des oxyfuel Kalzinators von 0,95 m^3 m^{-3} vor der Aufreinigung und Kompression (CPU) angenommen werden kann. Als Nebenkomponente beinhaltet das Abgas hauptsächlich O_2. Wegen der großen Menge an Calcium im Kalzinator werden schwefelhaltige Gaskomponenten quasi vollständig abgeschieden, gleichzeitig aber auch die Bildung von Stickoxiden gefördert. Die NO_x-Emissionen stiegen mit zunehmendem Restsauerstoff an und lagen im Bereich von $500 \cdot 10^{-6}$ m^3 m^{-3} bis $750 \cdot 10^{-6}$ m^3 m^{-3}, sodass eine Abscheidung während des Kompressions- und Aufreinigungsschritts nötig wird.

Contents

List of symbols

Latin alphabet

Symbol	Unit	Meaning
A	m^2	Surface area
d_p	μm	Particle diameter
E_A	$kJ\,mol^{-1}$	Activation energy
E_{calc}	$mol\,mol^{-1}$	Calcination/regeneration efficiency
$E_{CO_2,carb}$	$mol\,mol^{-1}$	CO_2 capture efficiency of the carbonator
$E_{SO_2,calc}$	$mol\,mol^{-1}$	SO_2 capture efficiency of the calciner
e_{SPECCA}	$MJ\,t^{-1}$	Specific primary energy consumption for CO_2 avoided
f_{active}		Active particle fraction
f_{carb}	$mol\,mol^{-1}$	Fractional carbonation conversion
f_{calc}	$mol\,mol^{-1}$	Fractional calcination conversion
h_j	m	Height of reactor j
H_i	$MJ\,kg^{-1}$	Net calorific value
k		Sorbent deactivation constant
k_{carb}	s^{-1}	Carbonation reaction constant
k_{calc}	s^{-1}	Calcination reaction constant
M_i	kg	Mass of component i
$\dot{M}_i$	$kg\,h^{-1}$	Mass flow of component i
$\tilde{M}_i$	$kg\,kmol^{-1}$	Molar mass of component i
m	$mol\,mol^{-1}$	Inclination of linear function
$\dot{m}_i$	$kg\,m^{-2}\,s^{-1}$	Mass flux of component i
N_i	mol	Molar amount of component i
$\dot{N}_i$	$mol\,s^{-1}$	Molar flow of component i

(Continued on the next page)

Symbol	Unit	Meaning
n		Reaction order
n_{cycle}		Number of calcination and carbonation cycles
n_{age}		Performed number of calcination and carbonation cycles
p^*	bar	Normalising pressure (logarithmic calculation)
p_i	bar	Partial pressure of component i
Q_3	$m^3\,m^{-3}$	Cumulative volumetric particle size distribution
q_3	$m^3\,m^{-3}\,\mu m^{-1}$	Volumetric particle size density
R	$J\,mol^{-1}\,K^{-1}$	Universal gas constant
r_j	$mol\,s^{-1}$	Reaction rate of reaction j
$\bar{r}_j$	$mol\,s^{-1}$	Average reaction rate of reaction j
r^*	$mol\,mol^{-1}\,s^{-1}$	Normalising reaction rate (logarithmic calculation)
T	°C / K	Temperature
t	s	Time
t_{accu}	s	Time required to fill the solid flow measurement section
t_j^*	s	Critical reaction time of reaction j
V_{accu}	m^3	Accumulation volume of the solid flow measurement section
$\dot{V}_i$	$m^3\,h^{-1}$	Volume flow of component i
$W_{s,j}$	$kg\,m^{-2}$	Specific solid inventory of reactor j
y_i	$m^3\,m^{-3}$	Volume fraction of component i
$\bar{y}_i$	$m^3\,m^{-3}$	Logarithmic average concentration of component i
x_i	$kg\,kg^{-1}$	Mass fraction of component i

Greek alphabet

Symbol	Unit	Meaning
Δ	var.	Difference
ΔH_R	$kJ\,mol^{-1}$	Reaction enthalpy
η_{comb}	$kg\,kg^{-1}$	Combustion efficiency
η_{NO_x}	$mol\,mol^{-1}$	NO_x conversion
$\rho_{solid,bulk}$	$kg\,m^{-3}$	Bulk density of circulating solids
μ_{O_2}	$kg\,kg^{-1}$	Specific oxygen demand for combustion
μ_{CLK}	$kg\,kg^{-1}$	Clinker cement ratio
$\nu_{CO_2,i}$	$m^3\,h^{-1}\,m^{-3}\,h$	Volume flow ratio of process j on total CO_2
$\nu_{recycle}$	$m^3\,h^{-1}\,m^{-3}\,h$	Flue gas recirculation ratio
ξ_0	$mol\,s^{-1}\,mol^{-1}\,s$	Share of sorbent make-up on molar carbonate flow fed to the calciner
ξ_{IL}	$mol\,s^{-1}\,mol^{-1}\,s$	Integration level
ξ_{LR}	$mol\,s^{-1}\,mol^{-1}\,s$	Looping ratio
$\xi_{LR,active}$	$mol\,s^{-1}\,mol^{-1}\,s$	Active looping ratio
ξ_{MR}	$mol\,s^{-1}\,mol^{-1}\,s$	Make-up ratio
$\tau_{solid,j}$	s	Solid residence time of reactor j
$\tau_{calc,active}$	s	Active space time of calciner
$\tau_{carb,active}$	s	Active space time of carbonator
φ		Gas-solid contact factor
X	$mol\,mol^{-1}$	Molar content
X_{calc}	$mol\,mol^{-1}$	Molar calcination degree
X_{CO_2}	$mol\,mol^{-1}$	CO_2 carrying capacity
$X_{CO_2,max}$	$mol\,mol^{-1}$	Maximum CO_2 carrying capacity
$X_{CO_2,avg}$	$mol\,mol^{-1}$	Average CO_2 carrying capacity
X_r	$mol\,mol^{-1}$	Residual CO_2 carrying capacity
X_1	$mol\,mol^{-1}$	CO_2 carrying capacity in the 1st cycle
X_{recarb}	$mol\,mol^{-1}$	Recarbonation degree

List of subscripts

Index	Meaning
0	Initial (analysis)
0	Make-up (sorbent)
avg	Average
(g)	Gaseous
attr	Attrition
CaL	Calcium looping
CSTR	Continuously stirred tank reactor
carb	Carbonator / carbonation
calc	Calciner / calcination
clinker	Clinker
coal	Coal
dehyd	Dehydration
dry	Dry
EF	Ref. status: entrained flow calcination (TGA analysis)
eq.	Equilibrium
flue gas	Flue gas
heat-up	Analysis step: heat up (TGA analysis)
in	In /entering
loop	Circulating
loss	Loss (sorbent)
measured	Measured
N_2-free	Nitrogen free basis
NG	Natural gas
others	Others (trace elements)

(Continued on the next page)

Index	Meaning
out	Out / exiting
purge	Purge (sorbent)
r	Residual (Dispersion/Surface area)
raw meal	Ref. status: raw meal (TGA analysis)
reactor	Reactor
recycle	Recycle
solved	Solved/calculated
sample	Sample (TGA analysis)
tot	Total

List of elements and molecules

Formula	Meaning
Al_2O_3	Aluminium oxide
C	Carbon
CO	Carbon monoxide
CO_2	Carbon dioxide
Ca	Calcium
Ca^{2+}	Calcium (ionic)
CaO	Calcium oxide
$CaCO_3$	Calcium carbonate
$CaSO_4$	Calcium sulfate
Ca_2SiO_4	Dicalcium silicate (belite)
Fe_2O_3	Ferric oxide
H	Hydrogen (atomic)
H_2O	Water (steam)
MgO	Magnesium oxide
N	Nitrogen (atomic)
N_2	Nitrogen (molecular)
NO_x	Nitrogen oxides
N_2O	Nitrous oxide
O	Oxygen (atomic)
O_2	Oxygen (molecular)
O^{2-}	Oxygen (ionic)
S	Sulphur
SO_2	Sulphur dioxide
Si	Silicon
SiO_2	Silicon dioxide

List of acronyms

Acronym	Meaning
0D	Zero dimensional (model)
1.5D	Axial model including particle distribution and reactions
2D-scenario	Scenario in which the average global temperature rise is limited to 2 °C
ASU	Air separation unit
ad	Air dried
BECCS	Bio energy carbon capture and storage
BET	Brunauer-Emmett-Teller (theory)
BFB	Bubbling fluidised bed
CaL	Calcium looping
CCS	Carbon capture and storage
CCU	Carbon capture and utilisation
CEMCAP	Project acronym - CO_2 capture from cement production
CFB	Circulating fluidised bed
CFD	Computational fluid dynamics
CLEANKER	Project acronym - CLEAN clinKER by calcium looping for low-CO_2 cement
CPU	CO_2 processing/purification unit
CSTR	Continuous stirred tank reactor
DIVA	Acronym of entrained flow reactor
DeNOx	Flue gas denitrification
G	Geseke (raw meal)
ID fan	Induced draft fan
IEA	International Energy Agency
IFK	Institute of Combustion and Power plant Technology

(Continued on the next page)

Acronym	Meaning
IPCC	Intergovernmental Panel on Climate Change
MAGNUS	Acronym of fluidised bed pilot facility
MFC	Mass flow controller
NDIR	Non-dispersive infrared
NRMSD	Root-mean-square deviation
PSA	Pressure swing absorption
PTFE	Polytetrafluoroethylene
R	Rumelange (raw meal)
TGA	Thermogravimetric analysis
V	Vernasca (raw meal)
wf	Water free
waf	Water and ash free
XRD	X-ray diffraction

List of figures

List of tables

Chapter 1

Introduction

1.1 CO₂ capture from cement plants

To mitigate anthropogenic driven climate change international institutions, such as the IPCC or IEA, propose besides other measures the implementation of carbon capture and storage technologies to reduce industrial CO_2 emissions [95, 97]. The cement sector is one of the largest industrial CO_2 emitters, emitting annually approx. 2.2 Gt of CO_2 which corresponds to around 7 % of the global anthropogenic CO_2 emissions [95, 97]. In contrast to other energy intensive industries such as power generation or steel production, the cement sector cannot be decarbonised by switching to alternative energy sources such as 'biogenic fuels', 'renewable electricity' or 'green hydrogen', since a major share of a cement plant's CO_2 emissions are inherent to its cement clinker production [95, 150]. Around two thirds of the cement plant's CO_2 emission originate from the calcination of carbonates present in the cement raw meal (i.e. feedstock), whereas the remaining third is attributed to the actual clinker burning process [150]. Consequently, either post-combustion or oxy-fuel CCS technologies need to be applied to decarbonise the cement sector [95, 150]. In the 2D-scenario of the IEA, CCS technologies are expected to contribute to cumulative CO_2 savings of 48 % until 2050, while improving the cement plant's energy efficiency should contribute additional 3 %. Switching to alternative, less carbon intensive fuels is supposed to save 12 % and reducing the required amount of clinker to form concrete (i.e. clinker to cement ratio) is expected to contribute to a cumulative reduction of 37 %. Electrification of the clinker production employing calcination by electrolysis [61] or heat supply using plasma torches and microwave heating [181] are currently under development. However, given the early stages of their development and the increasing demand for cement driven by the growth of emerging countries, these alternative production pathways are unlikely to allow substantial CO_2 emission reduction from the cement sector in the foreseeable future. Furthermore, all production pathways will inevitably emit CO_2 during calcination requiring subsequent sequestration or utilisation of the generated CO_2.

1.2   Motivation and objectives

Calcium looping is an emerging second generation post-combustion CCS technology which was originally proposed to decarbonise fossil-fuelled power plants. CO_2 is separated from a flue gas and subsequently transferred into a CO_2 rich gas stream by cyclic calcination (i.e. sorbent regeneration) and carbonation (i.e. sorbent loading) of a calcium containing solid sorbent. The calcium looping CO_2 capture process is characterised by a high energy efficiency owed to its high exergy level allowing efficient heat recuperation of the energy required for sorbent regeneration.

The calcium looping technology appears to be especially suitable for an application in the cement sector since both the calcium looping process and the cement clinker manufacturing process utilise $CaCO_3$ as common feedstock. Their inherent utilisation of $CaCO_3$ enables the reutilisation of spent sorbent from the calcium looping process within the clinker manufacturing process enabling higher sorbent make-up rates that eventually enhance the sorbent's CO_2 capture properties. Furthermore, synergies in terms of heat or energy integration can be raised increasing the energy efficiency of the entire system. Besides, calcium looping CO_2 capture from cement plants allows to generate net negative CO_2 emission by firing biogenic fuels or fuels with high biogenic share, such as municipal waste or sewage sludge, in the cement plant's pre-calciner and the calcium looping process. Such bio-energy carbon capture processes (BECCS) are expected to be required to counteract an overshot of CO_2 emissions [97].

So far, the calcium looping CO_2 capture process has been extensively studied for the decarbonisation of fossil-fuelled power plants, whereas the integration and implementation of calcium looping CO_2 capture into cement plants has mostly been assessed by means of process simulation with generic assumptions for instance that calcium looping purge (i.e. CaO) from a nearby power plant application replaces the $CaCO_3$ of the cement raw meal [141, 143]. Little experimental data is available for calcium looping CO_2 capture at operation conditions representing those anticipated for cement plant application. Therefore, this thesis aims to close this lack of experimental data by a comprehensive experimental assessment of calcium looping CO_2 capture for cement plants. The key objectives of this work are as listed below:

1. The development of various integration options suitable for a large variety of cement plants depending on the local boundary conditions. Two different reactor concepts are envisaged namely fluidised bed as well as entrained flow reactors.

2. Investigation of a wide range of potential calcium looping sorbents including limestones, mixed raw meals as well as raw meal components regarding their CO_2 capture properties using thermogravimetric analysis (TGA).

3. Experimental assessment and demonstration of the fluidised bed calcium looping technology at semi-industrial scale using University of Stuttgart's fluidised bed pilot plant addressing the boundary conditions identified within the development of the fluidised bed integration options.

4. Assessment of the sorbent performance for an entrained flow calcium looping system

The work conducted within the framework of this thesis is partly part of two European research projects, *CEMCAP* and *CLEANKER*. The obtained results have been shared within the consortia to validate and calibrate developed models.

1.3 Previously published results

The experiments presented within the framework of this thesis have been conducted with the support of colleagues of the department Decentralised Energy Conversion of the Institute of Combustion and Power Plant Technology (IFK) and various student theses which have been supervised by the author of this dissertation [31, 36, 101, 117, 159, 169, 182]. All results included in the student theses have been carefully assessed and the corresponding data and analyses have been reevaluated by the author to ensure consistent evaluation of the presented results. Previously published results and text passages of the author's own creation are included in this dissertation. Intermediate results have been published at conferences [87–90] and workshops [84, 91, 92]. Furthermore, the results of the investigation of the fluidised bed calcium looping CO_2 capture from cement plants (chapter 6) have partially been published in two peer-reviewed journal articles [85, 86].

Chapter 2

State of the art

2.1 Cement clinker manufacturing process

The main component of cement is cement clinker which is blended with other components such as granulated slag, silica dust, natural or industrial puzzolans, fly ash, burned shale and limestone [113]. Generally, different cement types are classified by DIN EN 197-1 [72] based on their composition mainly the clinker share in combination with clinker substitutes and the amount of the various clinker phases.

The main phases of cement clinker are tricalcium silicate (alite), dicalcium silicate (belite), tricalcium aluminate and calcium aluminoferrite [114, 160, 172]. Alite and belite contribute mainly to the overall strength of the concrete. Furthermore, alite contributes to the long-term and the initial strength of the concrete since it reacts rapidly with water, whereas belite is less reactive and contributes to the concrete's strength at later stages. Tricalcium aluminate contributes to the initial strength by a fast reaction and calcium aluminoferrite is responsible for the concrete's colour and its sulfate resistance .

Depending on the raw meal humidity the clinker manufacturing process is either classified as a wet, semi-wet or dry process. All processes have in common, that raw meal is preheated in a suspension preheater (cyclone preheater) by counter flowing flue gas. Depending on the raw meal humidity and the available sensible heat, three to six preheater stages are deployed [113, 150]. At each stage raw meal is injected into the riser duct of the subsequent cyclone and is entrained by the hot flue gas. Within several seconds the temperature of the flue gas and raw meal equilibrate and the two phases are separated by a cyclone. Subsequently, the preheated raw meal is calcined in a pre-calciner at temperatures of approx. 860 °C. The energy required for calcination is provided by the combustion of a fuel. The pre-calcined raw meal is then fed to the rotary kiln where clinker phases are formed at temperatures up to 1450 °C. The raw meal is slowly transported towards the rotary kiln's exit by its rotation and inclination (2.5 - 4.2 %) [150]. Within the rotary kiln, the raw meal partially melts and thereby, facilitating the solid-solid reactions that form the clinker phases. The formed clinker phases are stabilised by rapid cooling of the clinker in a clinker cooler. The heated cooler air is further utilised as combustion gas in the rotary kiln (secondary air) and in the pre-calciner (tertiary air). CO_2 emissions are inherent to the clinker production. Around two thirds of a cement plant's CO_2 emissions originate from raw meal calcination, whereas the remaining third is attributed to the actual clinker burning process [150]. Specific CO_2 emission depend on the burnt fuel but range typically around 850 kg of CO_2 per tonne of clinker [17, 150].

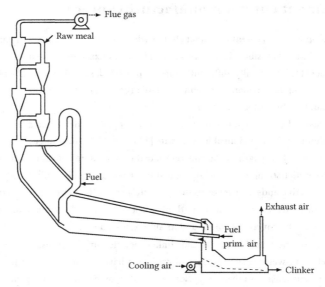

Figure 2.1: Clinker manufacturing process according to the best available techniques [150].

2.2   Carbon capture and storage technologies

CCS technologies are usually classified by three different approaches namely pre-combustion, oxy-fuel and post-combustion CO_2 capture technologies (figure 2.2). The concept of pre-combustion technologies is to generate a carbon-free fuel which is then burned or

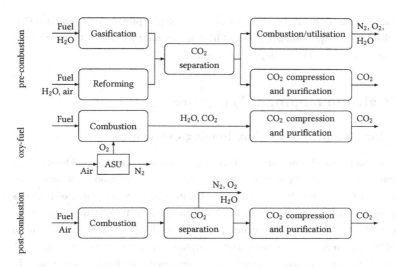

Figure 2.2: Overview of carbon capture technologies.

utilised in the downstream process generating a carbon-free flue gas. Most applications rely on fuel reforming or gasification in combination with a CO_2 separation technology to generate the carbon-free fuel that can be burnt or used in the subsequent process. In contrast, oxy-fuel technologies are based on the combustion of carbonaceous fuels in the absence of nitrogen to generate a flue gas which consists mainly of CO_2, steam and excess oxygen. Fuel is combusted with a mixture of recirculated flue gas and oxygen to adjust the combustion temperature and maintain characteristic heat transfer properties. Post-combustion CO_2 capture technologies separate CO_2 from the flue gas after the combustion of carbonaceous fuels. Most processes separate CO_2 from the flue gas utilising solvents or sorbents. The solvent or sorbent binds the CO_2 physically or chemically in one reactor (sorbent loading/absorber) and releases the CO_2 in a second reactor (sorbent regeneration/desorber). The capture and regeneration operation conditions are specified by the physical or chemical equilibrium of the utilised reaction system.

A second type of post combustion CO_2 capture technologies, such as membrane separation or cryogenic rectification, exploit different physical properties of the flue gas components such as diffusivity, permeability or dew point to separate CO_2 from the other flue gas components. Generally, the performance of the separation processes improves with increasing CO_2 concentration in the treated flue gas due to an increased driving force enhancing the absorption reaction or diffusion [180]. All CCS technologies provide a concentrated CO_2 stream that can be sequestrated (CCS) or utilised for various processes (CCU) after being compressed and liquefied in a CO_2 processing unit (CPU).

2.3 Calcium looping CO_2 capture

2.3.1 Development of calcium looping CO_2 capture

The idea to use lime or dolomite (i.e. CaO) as CO_2 sorbent was first proposed back in 1967 by Curran et al. [46] aiming to enhance the gasification of bituminous coal char in the CO_2 Acceptor Gasification Process. Subsequently, the carbonation reaction of CaO was investigated by Barker [24] and Bhatia and Perlmutter [29]. In 1999, Shimizu et al. [155] further developed the idea of the acceptor process to a full CCS technology applying an oxy-fuel fired calciner as regeneration and desorption reactor. In 2005, Abanades et al. [5] proposed various derivates of the calcium looping reactor configuration such as indirect heat transfer to the calciner or in-situ CO_2 capture in a fluidised bed boiler. The sorbent performance of lime and dolomite during the repetitive calcination and carbonation cycles was broadly investigated using thermometric analysis by various research groups [3, 24, 46, 76, 77, 145, 155, 157, 162, 176]. Process demonstration activities initially started with batch experiments using electrically heated fluidised bed reactors operating in various fluidisation regimes (bubbling bed [4, 145], fast fluidisation [116], circulating fluidised bed [2]). Subsequently, continuous operation was demonstrated applying different dual fluidised bed reactor concepts such as CFB-CFB configurations [138] or BFB-CFB configurations [39, 59, 60]. First non-electrically heated calcium looping systems in the range of 0.2 MW to 1.7 MW were erected around 2010 and successfully demonstrated CO_2 capture from synthetically mixed and real flue gases since then [10, 21, 22, 50, 52, 54, 55, 81, 82].

Concurrently to the experimental investigation, simulation work progressed [53]. Reaction models based on grain size models [167] or random pore models [29, 75] were deployed to describe the carbonation [13, 80, 104, 142, 184], sorbent sulphation [139, 184] and calcination [62, 183, 184] reactions in various simulation models [125]. These models ranged from 0D models [13, 124, 137] over 1.5D models [104, 139, 183, 184] (often based on the Kunii and Levenspiel [102] fluidized bed hydrodynamic model) to full CFD models [130, 131, 183]. Furthermore, the CO_2 capture cost, process integration into the upstream power plant, CO_2 purification and supply of oxygen was assessed by process simulation [63, 144].

Among other topics, research on calcium looping CO_2 capture focuses currently on the decarbonisation of other CO_2 emitters such as the cement industry (as presented in this work) [19, 85, 86] or on enabling flexible operation to allow load following of fossil fuelled power plants [51, 127–129].

2.3.2 The calcium looping CO_2 capture process

The principle of the calcium looping CO_2 capture process is based on the CaO-CO_2-$CaCO_3$ reaction system which can be described by equation 2.1. CO_2 is separated from a flue gas by cyclic calcination and carbonation of $CaCO_3$ or CaO, respectively.

$$CaO + CO_2 \rightleftharpoons CaCO_3 \qquad (2.1)$$

A simplified schematic of the calcium looping process is depicted in figure 2.3. The CO_2 capture process consists of an absorption reactor, called carbonator, and a desorption reactor, called calciner, which are interconnected by circulating sorbent. In the carbonator, CO_2 is captured from the flue gas by carbonising CaO. The CO_2 depleted flue gas is vented from the carbonator to the atmosphere and the carbonised CaO or loaded sorbent (i.e. $CaCO_3$) is transferred into the calciner. There, the loaded sorbent is regenerated (i.e. calcined) and the released CO_2 is sent to the CO_2 processing unit (CPU), whereas the regenerated sorbent (i.e. CaO) is fed back to the carbonator. During the repetitive calcination and carbonation cycles the sorbents ability to bind CO_2 decreases. Therefore, spent sorbent must be purged and replaced with a make-up of fresh sorbent.

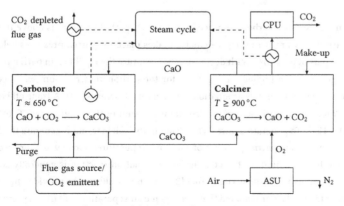

Figure 2.3: Schematic of the calcium looping CO_2 capture process.

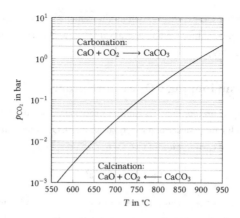

Figure 2.4: Calcination carbonation equilibrium of the CaO-CO_2-$CaCO_3$ reaction system derived from the FactPS database [23].

The operation conditions of both reactors are defined by the equilibrium of the calcination-carbonation reaction system. The equilibrium CO_2 partial pressure can be described by equation 2.2 in the temperature range from 500 °C to 1000 °C using the universal gas constant (R) and an activation energy (E_A) of 169.5 kJ mol^{-1} and is further depicted in figure 2.4. The equation was derived from FactSage database 'PureSolids' [23].

$$p_{CO_2,eq}(T) = 4.0009 \cdot 10^7 \text{bar} \cdot \exp\left(\frac{-E_A}{R \cdot T}\right) \tag{2.2}$$

At a given temperature carbonation occurs if the CO_2 partial pressure is above the equilibrium partial pressure. Contrary, calcination takes place for partial pressures below the equilibrium partial pressure. With decreasing temperature the equilibrium partial pressure decreases generating a higher driving force for the carbonation reaction, while concurrently the rate constant of the carbonation reaction rate decreases [29]. A trade off between pressure gradient and carbonation rate constant is found in the range of 600 °C to 700 °C [29, 74, 116, 156]. To generate a pure CO_2 stream in the calciner and regenerate the loaded sorbent an equilibrium partial pressure of 1 bar or temperatures above 900 °C are required. The reaction heat required for the endothermic calcination reaction is generally supplied by oxy-fuel combustion of an auxiliary fuel [30, 155] but could also be provided by indirect heat transfer [5, 99]. It is anticipated that a cryogenic air separation unit (ASU) provides the required oxygen for oxy-fuel combustion since the oxygen demand exceeds the capacities of state of the art membrane or PSA systems [144]. The total amount of oxygen required for a calcium looping system is significantly reduced compared to oxy-fuel systems in the energy sector as only a fraction of the fuel is burned at oxy-fuel conditions. The high exergy level of the calcium looping process enhances and necessitates energy recuperation. Generally, heat can be recuperated at three different locations: (i) the reaction heat of the exothermic carbonation reaction in the carbonator, (ii) sensible heat from the solids exiting the calciner and (iii) sensible heat from the gas streams exiting the calciner and carbonator. Sufficient energy can be recuperated from these locations to drive an efficient state of the art steam cycle [105, 106, 135, 140, 142]. De Lena et al. [105] concluded that an industrial heat recovery steam cycle with regenerative feed-water heating is most suitable for cement plant application due to the fact that (i) energy recuperated from the calcium looping system cannot be integrated into the steam cycle of the power plant, (ii) 90 % of the heat can be recovered at temperatures above 320 °C and (iii) expensive materials required to raise the steam parameters can be avoided. Consequently, a subcritical steam cycle with an electric efficiency of approx. 35 % can be expected for a calcium looping CO_2 capture system in cement plants [105].

Key performance indicators and parameters to describe and assess the calcium looping CO_2 capture process are the looping ratio, the sorbent make-up ratio, the CO_2 capture efficiency as well as the CO_2 avoidance cost. The CO_2 capture efficiency ($E_{CO_2,carb}$) generally refers to the carbonator and describes the ratio between the absorbed amount of CO_2 in the carbonator and its CO_2 load (eq. 2.3). In order to account for different carbonation temperatures or the achievable CO_2 concentration, respectively, the carbonator CO_2 capture efficiency is often normalized with respect to the CO_2 capture efficiency that could be achieved if an equilibrium state is reached (i.e. equilibrium CO_2 capture). The equilibrium CO_2 capture efficiency ($E_{CO_2,carb,eq}$) can be calculated by equation 2.4 accounting for the reduction in volume flow associated with CO_2 absorption using the carbonator inlet CO_2 concentration ($y_{CO_2,carb,in}$) and the equilibrium CO_2 concentration ($y_{CO_2,eq}$). The application of the calcium looping technology in the cement sector enables the reutilisation of the calcium looping purge and substituting a part of the cement raw meal's calcium carbonate content. Consequently, the CO_2 emissions of the clinker manufacturing process are reduced. In addition, the sorbent make-up's calcination effort in the calcium looping calciner can be regarded as benefit contributing to the overall CO_2 capture of the calcium looping system as opposed to power plant application. Thus, the CO_2 capture efficiency of the overall system ($E_{CO_2,CaL}$) can be expressed by equation 2.5 neglecting fuel required for make-up calcination.

$$E_{CO_2,carb} = \frac{\dot{N}_{CO_2,carb,in} - \dot{N}_{CO_2,carb,out}}{\dot{N}_{CO_2,carb,in}} \tag{2.3}$$

$$E_{CO_2,carb,eq} = \frac{y_{CO_2,carb,in} - y_{CO_2,eq} \cdot \dfrac{1 - y_{CO_2,carb,in}}{1 - y_{CO_2,eq}}}{y_{CO_2,carb,in}} \tag{2.4}$$

$$E_{CO_2,CaL} = 1 - \frac{\dot{N}_{CO_2,carb,out}}{\dot{N}_{CO_2,carb,in} + \dot{N}_{CaCO_3,0}} \tag{2.5}$$

The looping ratio (ξ_{LR}) is defined as molar ratio of CaO transferred into the carbonator ($\dot{N}_{CaO,loop}$) with respect to CO_2 entering the carbonator ($\dot{N}_{CO_2,carb,in}$) (eq. 2.6). Therefore, it represents a measure of how much CaO can react with CO_2 in the carbonator as well as an indication on the required heating duty and heat recuperation system as the circulating flow of sorbent needs to be heated from carbonation to calcination temperature and cooled down vice versa.

$$\xi_{LR} = \frac{\dot{N}_{CaO,loop}}{\dot{N}_{CO_2,carb,in}} \tag{2.6}$$

Similarly, the make up ratio (ξ_{MR}) is defined as molar flow of fresh sorbent ($\dot{N}_{CaCO_3,0}$) being fed to the calcium looping system with respect to the amount of CO_2 entering the carbonator (eq. 2.7). With increasing sorbent make-up the residence time of the sorbent in the calcium looping system decreases. Consequently, the sorbent undergoes less calcination and carbonation cycles resulting in a higher CO_2 capture capacity.

$$\xi_{MR} = \frac{\dot{N}_{CaCO_3,0}}{\dot{N}_{CO_2,carb,in}} \tag{2.7}$$

The CO_2 capture effort can be expressed in various ways. Usually it is expressed in terms of energy required to avoid CO_2 emissions or in terms of costs required to avoid CO_2 emissions. To compare different capture technologies that use various types of energy the so called specific cost of CO_2 avoided (e_{SPECCA}) has been defined [140]. This approach calculates the primary energy demand of a capture technology based on its direct and indirect fuel consumption. Indirect fuel consumption associated to the use of electricity is rated by an average electricity generation efficiency [106, 140, 174]. An unique feature concerning calcium looping CO_2 capture is the fact that electricity can be generated by heat recuperation. This generated electricity offsets the primary energy demand accordingly.

2.3.3 Carbonation reaction

It is generally agreed that the carbonation of CaO follows a two stage reaction pattern [1, 3, 24, 29, 74, 78, 94]. An initial fast reaction regime in which the reaction rate is kinetically controlled is followed by a sharp transition to a slower reaction regime in which the reaction is limited by diffusion through the forming product layer.

The schematic reaction pattern is depicted in figure 2.5 as normalised conversion ($\frac{X_{CO_2}(t_{carb})}{X_{CO_2,max}}$) against the reaction time (t_{carb}). Bhatia and Perlmutter [29] investigated the carbonation of CaO after one calcination using thermogravimetric analysis (TGA) and found that the reaction follows a sigmoid shaped reaction profile. After a nucleation state to a conversion of approx. $0.1\,mol\,mol^{-1}$, the maximum reaction rate is reached followed by a sharp transition to the diffusion controlled regime. The same reaction profile was found for the carbonation of higher cycled sorbent up to 500 calcination and carbonation cycles, although the overall conversion is reduced due to the deactivation of the sorbent [74, 145]. Alonso et al. [9] highlighted that the transition's sharpness and transition time depends on the experimental set up. If the carbonation is limited by mass transfer or kinetics, the transition's sharpness is softened and the point of transition is moved to higher reaction times. However, the maximum achievable conversion is not affected by the carbonation reaction rate [3]. Generally, only the initial fast reaction regime is considered relevant for CO_2 capture, since exploiting the diffusion controlled reaction will require prolonged solid residence times in the carbonator and thus, enlarging the carbonator design disproportionately [3].

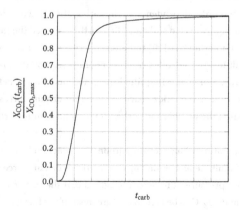

Figure 2.5: Schematic course of carbonation conversion presented as normalised conversion ($\frac{X_{CO_2}(t_{carb})}{X_{CO_2,max}}$) vs. carbonation reaction time (t_{carb}) generated from own TGA measurement.

Many authors [3, 13, 14, 24, 29, 35] concluded that the reaction is governed by the accessibility of the available surface area. During the kinetically controlled reaction step, the carbonation rate is controlled by smaller pores with a high specific surface area. Once these pores are blocked by the forming product layer (i.e. $CaCO_3$), the carbonation reaction is controlled by the reaction in larger pores which are still accessible but have a smaller specific surface area. Hence, the reaction rate declines. Finally, the carbonation reaction is controlled by the diffusion through the product layer once the larger pores are blocked. Alvarez and Abanades [14] found a product layer of approx. 49 nm to be characteristic for the transition from the chemical controlled regime to the diffusion controlled regime using a cylindrical pore model, whereas Bouquet et al. [35] proposed that the transition occurs at a product layer of 43 nm once the voids at micro grain level are filled. Barker [24] calculated a characteristic value of 22 nm based on the different molar volumes of CaO and $CaCO_3$ and the total surface available for the reaction [35]. Bhatia and Perlmutter [29] further concluded that limestones with wider pores react more slowly but have an overall higher carbonation conversion.

The dependency of the carbonation reaction rate from the operation conditions such as carbonation temperature and CO_2 partial pressure as well as particle size has mostly been assessed by TGA experiments [29, 74, 165, 185]. Bathia and Perlmutter [29] found a first order dependency for the fast reaction regime with respect to the CO_2 partial pressure and a zero order for the diffusion controlled reaction regime applying a random pore model to their TGA experiments investigating the first carbonation of calcined limestone. Grasa et al. [74] also found a first order dependency of the carbonation reaction in the fast reaction regime up to 1 bar for highly cycled sorbent. They further adapted the random pore model proposed by Bathia and Perlmutter [29] to better describe the transition from the fast reaction regime to the diffusion controlled reaction regime based on the product layer that is formed [75]. Contrarily, Sun et al. [165] fitted the grain model of Szekely et al. [168] and found that the order of the carbonation reaction is one for driving forces (i.e. partial pressures difference towards the equilibrium) below 0.1 bar and zero for driving forces above 0.1 bar.

The carbonation temperature influences the carbonation reaction by enhancing the intrinsic reaction rate but simultaneously reducing the reaction's driving force due to an increase of the equilibrium CO_2 pressure. It is generally agreed that the carbonator should be operated between 600 °C and 700 °C to be able to capture most of the flue gas' CO_2 (i.e reach low CO_2 partial pressures) with a sufficiently fast reaction rate. For these operation conditions a minor influence of the carbonation temperature on the carbonation reaction rate was found [29, 74, 156]. Bathia and Perlmutter [29] investigated carbonation temperature from 400 °C to 725 °C and found a maximum carbonation rate at 655 °C with a CO_2 concentration of $0.1\,m^3\,m^{-3}$. Silaban and Harrison [156] reported a slight increase in reaction rate raising the carbonation temperature from 550 °C to 650 °C with a subsequent reduction of the carbonation rate due to a reduced driving force for higher carbonation temperatures. The reduction was less pronounced at elevated total pressure [156]. There is consensus that the particle size does not effect the carbonation reaction [29, 74] nor the sorbent's CO_2 carrying capacity [3, 76, 118]. Although minor diffusion resistance for larger particles were found in the initial carbonation cycles [29, 74], the influence diminishes with subsequent cycles. Grasa et al. [74] could not observe any influence of the particle size on the carbonation reaction after the 20th cycle. Manovic and Anthony [118] attributed this behaviour to enhanced CO_2 diffusion through the particle's porous structure that gradually develops with increasing number of cycles. Different reaction orders are reported for the carbonation reaction due to the application of different reaction models and/or reaction mechanisms. Usually, random pore models [29, 75, 164] or grain size models [35, 112, 161] as well as semi empirical models [4, 13, 80, 155] are applied to describe the carbonation reaction. In accordance with the variety of applied models multiple expressions are proposed to describe the carbonation reaction. Most of the reaction models propose a first order dependency on the CO_2 partial pressure or the pressure difference towards the equilibrium CO_2 partial pressure as obtained from the respective experiments. Grain size models generally propose a dependence of the carbonation reaction on the CO_2 uptake potential ($X_{CO_2,max} - X_{CaCO_3}$) by the power of two thirds (eq. 2.8), whereas the semi empiric models postulate a linear dependency. Assuming full calcination a simple empiric model can be described by equation 2.9 using the maximum CO_2 carrying capacity of the sorbent ($X_{CO_2,max}$). Besides, random pore models use an intrinsic carbonation rate assessed with an assumed pore structure function.

$$\left.\frac{dX}{dt}\right|_{carb} = k_{carb} \cdot (X_{CO_2,max} - X_{CaCO_3})^{\frac{2}{3}} \cdot (y_{CO_2} - y_{CO_2,eq.}) \qquad (2.8)$$

$$\left.\frac{dX}{dt}\right|_{carb} = k_{carb} \cdot X_{CO_2,max} \cdot (y_{CO_2} - y_{CO_2,eq.}) \qquad (2.9)$$

2.3.4 Calcination reaction

Although the calcination of limestone is studied over a wide range of operation conditions, there is no consensus about the reaction mechanism and hence, the reaction models and kinetic data proposed in the literature differ substantially. Chen et al. [40] reported (in agreement with Wang et al. [177, 178]) that the required time to complete calcination reduces with increasing calcination temperature as well as with decreasing CO_2 partial pressure, but did not specify a particular dependency of the two parameters. Chen et al. [40] studied the sorbent calcination using thermogravimetric analysis, whereas Wang et al. [177, 178] studied calcination in a continuous fluidised bed reactor at high CO_2 partial pressures. Borgwardt [32] investigated the calcination of two limestones with a particle size below 90 μm in differential and entrained flow conditions in pure nitrogen and concluded that the calcination rate only depends on the particle surface. He proposed a uniform reaction model comprising the calcination reaction constant (k_{calc}) and the particle's surface area (A_{CaCO_3}) (eq. 2.10). Supposably, an effect of temperature could not be determined due to the very high driving force as the experiments were operated in pure nitrogen. Fernandez et al. [66] investigated the calcination of various cement raw meals, limestones and marlstones with a particle size below 50 μm at elevated CO_2 concentrations up to $0.85 \, m^3 \, m^{-3}$ and found a good agreement with the calcination model of Borgwardt in case of air calcination conditions. They further introduced a factor to account for high CO_2 partial pressures and to describe the influence of the calcination temperature based on the equilibrium CO_2 partial pressure difference and the equilibrium constant of the calcination carbonation reaction according to the van't Hoff equation [66].

$$r_{calc} = k_{calc} \cdot A_{CaCO_3} \qquad (2.10)$$

Various research groups [47, 48, 93, 96, 123, 132] found a linear dependency of the calcination rate with respect to the CO_2 partial pressure difference towards the equilibrium partial pressure for small ($d_p < 75 \, \mu m$) and medium sized particles ($d_p < 2 \, mm$). Okunev et al. [132] investigated $CaCO_3$ particles with size cuts between 0.25 mm and 0.3 mm, 0.8 mm - 0.9 mm up to a calcination temperature of 800 °C and concluded that the effect of temperature on the calcination is negligible and that the transition from diffusion controlled to reaction controlled regime depends on the particle size. The equilibrium CO_2 partial pressure fraction ($p_{CO_2}/p_{CO_2,eq.}$) at which the transition occurs decreases with decreasing particle size. Martinez et al. [123] reported good agreement with the reaction mechanism proposed by the model of Fang et al. [62] (eq. 2.11) in their TGA study investigating various limestone sizes, calcination temperatures and CO_2 partial pressures. They further highlighted that the model of Fang is based on the 'gas solid reaction model with a moving boundary' by Szekely and Evans [167] and that the predictions match those of random pore models developed to describe sorbent calcination if no or moderate transport resistances occur [123].

$$r_{calc} = k_{calc} \cdot (1 - f_{calc})^{\frac{2}{3}} \cdot (y_{CO_2,eq} - y_{CO_2}) \qquad (2.11)$$

Rao et al. [136] found that heat transfer becomes important for particle sizes above 6 mm inhibiting the calcination reaction [68]. Although the calcination mechanism is not un-equivocally elucidated, it can be concluded that the calcination is influenced primarily by the CO_2 partial pressure and calcination temperature or the driving force towards the equilibrium, respectively. With increasing temperature or decreasing CO_2 partial pressure calcination is accelerated. Heat and mass transfer limitations can occur for larger particle sizes but the inhibiting effect diminishes with decreasing particle size.

Back in 1985, Borgward [32] highlighted that the reactivity of the produced CaO is affected by the calcination conditions influencing the grain size and specific surface area of the sorbent and concluded that large surface areas are obtained at low temperatures due to the reduced influence of sintering on the grain growth. Chen [40] found that the effect of temperature on the surface area is less severe for temperatures below 1000 °C in case of oxy-fuel conditions compared to air calcination conditions. Nonetheless, the overall surface area and pore volume is reduced for oxy-fuel calcination conditions. At a calcination temperature of 1000 °C the pore size distribution shifted from 25 Å to 200 Å averaging at around 43 Å for air calcination conditions to 40 Å to 300 Å averaging at around 85 Å for an oxy-fuel atmosphere containing $0.8 \, m^3 \, m^{-3} \, CO_2$ and $0.2 \, m^3 \, m^{-3} \, O_2$ [40]. A more comprehensive outline on how the sorbent activity is affected can be found in section 2.3.5.

2.3.5   Sorbent activity

This section addresses the sorbent's ability to capture CO_2 which is generally referred to as CO_2 carrying capacity or sorbent activity. In conformity with the stoichiometry of the carbonation reaction (eq. 2.1), the sorbent's CO_2 carrying capacity is presented on a molar basis in terms of mole of CO_2 per mole of CaO. The sorbent's ability to absorb CO_2 is mainly affected by sintering (i.e. grain growth), sulphation and belite formation in case of raw meal based sorbents. Additionally, loss of sorbent due to attrition and elutriation of fines can occur reducing the calcium looping system's CO_2 uptake potential. In order to maintain steady CO_2 capture a constant sorbent activity is required. Therefore, spent sorbent must continuously be purged and replaced by fresh sorbent make-up. The extent of required sorbent make-up depends on the loss of sorbent activity. How the sorbent activity is affected by the operation conditions as well as the various sorts of loss of sorbent activity are described subsequently.

Sorbent conversion

Sorbent conversion or rather the CO_2 carrying capacity is mainly influenced by the carbonation temperature and slightly influenced by the CO_2 partial pressure. It has been unanimously reported that the sorbent conversion improves with increasing carbonation temperature [29, 45, 109, 110, 120, 133]. Li et al. [110] increased the carbonation temperature at the end of the carbonation phase and yielded an additional increase of the CO_2 carrying capacity. They attributed this behaviour to the coalescence of contiguous product islands. Surface diffusion and coalescence of product islands is facilitated at higher carbonation temperatures yielding more accessible CaO and thus, increasing the overall CO_2 carrying capacity of the sorbent. Manovic and Blamey [120] further reported that the improvement of sorbent conversion decreases for carbonation temperatures above 650 °C. As a consequence, beneficial operation windows of the carbonator lie between 600 °C and 700 °C, bearing in mind that lower carbonation temperatures reduce the achievable minimum CO_2 concentration. Other parameters that enhance solid state diffusion such as steam or sodium content also promote higher sorbent conversions since they facilitate the coalescence of product islands similar to the effect of elevated temperatures [120]. The importance of this promoting effect decreases with increasing temperature as solids state diffusion increases overall with temperature [120]. Similarly, the effectiveness of a particular effect promoting solid state diffusion is interdependent with the other promoting effects.

The influence of the CO_2 partial pressure on the sorbent conversion is less unambiguous. Ortiz et al. [133] investigated various carbonation atmospheres with $0.15\,\mathrm{m}^3\,\mathrm{m}^{-3}$, $0.30\,\mathrm{m}^3\,\mathrm{m}^{-3}$, $0.45\,\mathrm{m}^3\,\mathrm{m}^{-3}$ and $0.60\,\mathrm{m}^3\,\mathrm{m}^{-3}$ of CO_2 using thermogravimetric analysis and concluded that the transition from the fast, reaction controlled regime to the diffusion controlled reaction regime is shifted to higher conversions with increasing CO_2 concentration while the overall conversion remained unaffected. Contradictorily, Bhatia and Perlmutter [29] investigated CO_2 partial pressures up to 0.42 bar at various carbonation temperatures (400 °C to 725 °C) and concluded that the carbonation conversion is not affected in the investigated range of parameters. Nevertheless, they both found a first order dependency of the reaction rate with respect to the CO_2 partial pressure. Besides, Grasa et al. [74, 76] reported reduced sorbent conversions for CO_2 partial pressures above 0.5 bar which they attributed to pore blockages by the forming product layer. As the sorbent becomes more porous with increasing number of cycles pore blockage was less inhibiting [76].

In conclusion, the CO_2 carrying capacity increases with enhanced solid state diffusivity namely the presence of steam as well as increasing carbonation temperatures, whereas the

CO_2 partial pressure scarcely affects the sorbent conversion but its conversion rate.

Cyclic sorbent activity

It is unanimously reported [1, 6, 24, 74, 76, 77, 145, 157, 162, 176], that during the repetitive calcination and carbonation cycles the sorbent's ability to bind CO_2 decreases exponentially towards an asymptotic threshold. The decay of sorbent activity of limestone based sorbents can be described for most limestones in good agreement by a semi-empirical approach proposed by Grasa et al. [76] (eq. 2.14). This model was developed following the general power law expression with a reaction order of two used to describe the deactivation of catalysts accounting for a residual dispersion or surface area (A_r/A_0) (eq. 2.12), respectively [25]. Assuming that the surface area is proportional to the sorbent activity the sorbent activity can be expressed by equation 2.13. Integration of equation 2.13 yields an expression to calculate the sorbent activity as a function of the number of calcination and carbonation cycles (n_{cycle}) based on the sorbent's residual activity (X_r) and its deactivation constant (k).

$$-\frac{d\frac{A}{A_0}}{dn_{cycle}} = k \left(\frac{A}{A_0} - \frac{A_r}{A_0}\right)^2 \tag{2.12}$$

$$-\frac{dX_{CO_2}}{dn_{cycle}} = k \left(X_{CO_2} - X_r\right)^2 \tag{2.13}$$

$$X_{CO_2}(n_{cycle}) = \frac{1}{\frac{1}{1-X_r} + k \cdot n_{cycle}} + X_r \tag{2.14}$$

Minor deviation in the deactivation constant (k) is observed depending on the calcination conditions. At more severe calcination conditions (i.e. higher temperatures, CO_2 partial pressure or prolonged residence times) an increased deactivation constant is observed [76]. Reported deactivation constants range from 0.48 to 1, whereas the residual sorbent activity range from $0.04\,mol\,mol^{-1}$ to $0.17\,mol\,mol^{-1}$. However, some limestone have a deactivation constant of up to 1.96. Cycle experiments yield similar results despite the comparatively large differences in experimental set-up, operation conditions and utilised sorbents. An excerpt of reported sorbent characteristics regarding their cyclic CO_2 carrying capacity is presented in table 2.1. Furthermore, figure 2.6 presents the cyclic CO_2 carrying capacity of various limestones derived from the work of Grasa et al. [74]. Grasa et al. [76] highlighted that the CO_2 carrying capacity of most sorbents can adequately be described using a deactivation constant of 0.52 and a residual sorbent activity of $0.075\,mol\,mol^{-1}$. Alonso et al. [9] reviewed and summarised the effect of various operation parameters on the CO_2 carrying capacity as follows. The influence of the calcination temperature is negligible

for temperatures below 950 °C [76], but becomes more pronounced at temperatures above
950 °C [73, 76]. The sorbent's particle size does not influence the CO_2 carrying capacity indicating that pore blockage does not prevent the carbonation of the particle's inner surface
[3, 29, 76, 118]. While larger particle sizes show a slightly reduced reaction rate, the reaction
rate does not affect the overall achieved CO_2 loading [3]. Also, the type of limestone has
minor effects on the deactivation curves, although some limestones hold a higher residual
CO_2 carrying capacity than others [16, 76, 162].

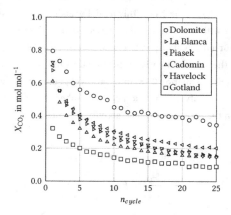

Figure 2.6: Sorbent CO_2 carrying capacity (X_{CO_2}) vs. calcination and carbonation cycles
(n_{cycle}), derived from Grasa et al. [74].

Sorbent sintering

From research related to flue gas desulphurisation using limestone based sorbents it is well
known that sintering of CaO occurs during calcination. Sintering is mainly affected by
the calcination temperature, the residence time, the CO_2 partial pressure, the presence of
steam in the reaction atmosphere and by impurities present in the sorbent [32–34, 67]. The
growth and coalescence of CaO grains lead to a reduction of the sorbent's specific surface
area and of its porosity. Consequently, the potential conversion with gaseous reactants
such as CO_2 or SO_2 decreases concurrently with the decreasing accessible surface area. It
is widely agreed that the sinter rate strongly increases with increasing calcination temperature and residence time [32, 34, 67]. Borgwardt [34] investigated among other sorbents
$CaCO_3$ derived from limestone and found that the specific surface area significantly de-

Table 2.1: Reported sorbent deactivation constants (k) and residual sorbent activities (X_r) for various sorbents at different operation conditions.

Author, Sorbent	k	X_r	n_{cycle}	Calcination conditions	Carbonation conditions
Barker [24]*,	0.33	0.04	10	T_{calc} = 866 °C	T_{carb} = 866 °C
Curran et al. [46]*, South Dakota Limestone	0.52	0.13	11	T_{calc} = 1050 °C	T_{carb} = 815 °C
Silaban et al. [156]*, Mallinckrodt	0.986	0.175	5	T_{calc} = 750 °C	T_{carb} = 750 °C
Shimizu et al. [155]*, Chichibu	1.23	0.167	4	T_{calc} = 950 °C	T_{carb} = 600 °C
Fennell et al. [64]*, Purbeck	0.40	0.113	19	T_{calc} = 950 °C	T_{carb} = 750 °C
Grasa et al. [76],				t_{calc} < 20 min	T_{carb} = 650 °C,
	0.52	0.075	N/A	T_{calc} = 850 °C	t_{carb} = 5 min,
La Blanca, 0.40 - 0.60 mm	0.52	0.075	150	T_{calc} = 900 °C	p_{CO_2} = 0.01 MPa
	0.52	0.075	150	T_{calc} = 950 °C	
Grasa et al. [76],				t_{calc} = 20 min	T_{carb} = 650 °C,
La Blanca, 0.40 - 0.60 mm	1.73	0.075	10	T_{calc} = 950 °C	t_{carb} = 5 min,
	1.73	0.075	10	T_{calc} = 1000 °C	p_{CO_2} = 0.01 MPa
Grasa et al. [76],					T_{carb} = 650 °C,
Dolomite	0.28	0.22	68	T_{calc} = 900 °C	t_{carb} = 10 min
	0.39	0.11	80	T_{calc} = 1000 °C	p_{CO_2} =
Grasa et al. [76],				T_{calc} = 950 °C	0.01 MPa
	0.48	0.075		t_{calc} = 2 min	T_{carb} = 650 °C,
	0.60	0.075	200	t_{calc} = 3 min	t_{carb} = 5 min,
La Blanca, 0.40 - 0.60 mm	0.76	0.075	120	t_{calc} = 10 min	p_{CO_2} = 0.01 MPa
	0.75	0.075	100	t_{calc} = 30 min	
	1.00	0.075	25	t_{calc} = 60 min	
Grasa et al. [76],					
La Blanca, 0.40 - 0.60 mm	0.52	0.075	> 100		T_{carb} = 650 °C,
Piasek, 0.40 - 0.60 mm	0.52	0.075	> 100	T_{calc} = 850 °C,	t_{carb} = 10 min,
Cadomin, 0.40 - 0.60 mm	0.52	0.075	> 100	t_{calc} = 10 min	p_{CO_2} = 0.01 MPa
Havelock, 0.40 - 0.60 mm	0.52	0.075	> 100		
Gotland, 0.40 - 0.60 mm	1.96	0.075	> 100		
Dolomite, 0.40 - 0.60 mm	0.28	0.22	> 100		
Grasa et al. [76],					T_{calc} = 650 °C,
La Blanca, 0.10 - 0.25 mm	0.52	0.075	40		
La Blanca, 0.25 - 0.40 mm	0.52	0.075	40	T_{calc} = 850 °C,	t_{carb} = 20 min,
La Blanca, 0.40 - 0.60 mm	0.52	0.075	40	t_{calc} = 10 min	p_{CO_2} = 0.01 MPa
La Blanca, 0.60 - 0.80 mm	0.52	0.075	40		
La Blanca, 0.80 - 1.00 mm	0.52	0.075	40		

* derived from reference figures

creased for calcination temperatures above 900 °C. The presence of either steam or CO_2 during calcination also increases the loss of surface area or porosity associated with sintering [33, 40]. Borgwardt [33] reported that the impact of steam is stronger than that of CO_2 and that their particular effect is additive. Furthermore, the improving effect of steam on sintering diminishes with increasing steam concentration. Borgward [33] concluded that the surface and grain boundary diffusion is enhanced if either one or both components are present during calcination. Similarly, impurities that improve solid state diffusion accelerate sintering [34].

With increasing numbers of calcination and carbonation cycles the sorbent develops a bimodal pore size distribution [15, 163]. Initial calcines show a single characteristic peak for small pore sizes below 100 nm which decreases with increasing number of calcination and carbonation cycles. Concurrently, a second peak develops at a larger pore diameter growing with increasing number of calcination and carbonation cycles. The small pore peak can be attributed to fresh calcined $CaCO_3$ formed by the previous carbonation, whereas the peak at larger pores originates from pore amalgamation associated with CaO sintering [15, 163]. This decreasing share of small pores on the overall pore volume and specific surface area, respectively, are in good agreement with the decreasing sorbent activity with increasing calcination and carbonation cycles [64]. Thus, it can be concluded that the surface area provided by the small pores is majorly contributing to the sorbent's CO_2 absorption capability.

Sorbent sulphation

SO_2 originating from the flue gas as well as from the combustion of the auxiliary fuel in the calcium looping calciner can react with the sorbent's calcium forming calcium sulfate. Calcium sulfate binds free lime, thus CaO is unable to react with CO_2 and sorbent deactivation is accelerated. Sorbent deactivation by sulphation is considered non-reversible since the decomposition temperature of $CaSO_4$ is above 1200 °C [166], which exceeds the calcination temperature by a large margin. Generally, gaseous SO_2 can react with the sorbent's calcium via two reaction pathways. Indirect sulphation (eq. 2.15) occurs at calcining conditions. $CaCO_3$ is first calcined and is subsequently sulphated. Contrarily, in non-calcining conditions sulphation takes place via the direct sulphation path way (eq. 2.16) [158]. Both reactions require excess oxygen to oxidise the calcium sulfit ($CaSO_3$) or SO_2 [18].

$$CaO + SO_2 + 0.5\,O_2 \longrightarrow CaSO_4 \tag{2.15}$$

$$CaCO_3 + SO_2 + 0.5\,O_2 \longrightarrow CaSO_4 + CO_2 \tag{2.16}$$

The additional reduction of sorbent CO_2 carrying capacity in the presence of SO_2 has consensually been reported and associated with the competition of SO_2 and CO_2 for calcium [77, 119, 121, 145]. Moreover, the SO_2 absorption performance of cycled sorbent proved to be enhanced, due to the shift towards a larger pore network at elevated cycles, which prevents plugging of pores by the forming $CaSO_4$ product layer [77, 119, 121]. Arias et al. [20] investigated calcium looping SO_2 co-capture and proved its feasibility at a 30 kW test facility. For desulphurisation of fluidised bed combustors, pore blockage by the forming $CaSO_4$ product layer is limiting the degree of sorbent utilisation. However, Grasa et al. [77] anticipated that the sorbent's sulphate conversion is not limited by pore blockage of $CaSO_4$ due to the high sorbent make-up rates required to maintain a constant CO_2 carrying capacity of the calcium looping system. Hence, sorbent sulphation constitutes an additional sorbent deactivation mechanism, but not a crucial one, since sorbent deactivation by sulphation can simply be counteracted by increasing the sorbent make-up. Simultaneously, SO_2 co-capture is enabled allowing to forgone an additional desulphurisation unit.

Sorbent attrition

The main sorbent attrition mechanisms for fluidised bed calcium looping systems are (i) primary fragmentation or decrepitation, (ii) secondary fragmentation and (iii) abrasion [146]. Decrepitation arises once the sorbent is injected into the calcium looping calciner due to the thermal stress resulting from the rapid heat up of the particle and due to internal pressure generated by the release of CO_2 (i.e. calcination). Primary fragmentation creates both coarse and fine particles [146, 147]. Secondary fragmentation as well as abrasion occurs due to mechanical stresses resulting from collision with other particles or the reactor wall. While secondary fragmentation generates coarse particles that are nonelutriable, abrasion produces fines which are discharged by elutriation [146, 147]. Due to the reduced particle interaction in entrained flow reactors, secondary fragmentation is supposed to sparsely contribute to the overall sorbent attrition. The highest elutriation rates are observed within the first calcination due to the large contribution of primary fragmentation. Subsequently, attrition rates decline since the sorbent is hardened due to sintering and sulphation (i.e. formation of a hard $CaSO_4$ shell) [42, 43]. With progressing calcination the sorbent becomes more porous and consequently more friable [43]. As a consequence, attrition during calcination exceeds attrition during carbonation [42]. Alonso et al. [8] investigated the attrition behaviour of four limestones in an electrically heated dual circulating fluidised bed system and found increased attrition during the first calcination due to decrepitation. The severity of decrepitation depended on the utilised limestone. They reported an increase of the attrition rate from $2 \% \, h^{-1}$ without calcination to $5.3 \% \, h^{-1}$, $11.4 \% \, h^{-1}$, $16.7 \% \, h^{-1}$, and $17.5 \% \, h^{-1}$, respectively, during calcination. Coppola et al. [42, 43] investigated attrition characteristics of six limestones at constant dry calcination and carbonation conditions as well as the influence of steam and SO_2 on the attrition behaviour of one limestone and reported elutriation rates between $0.3 \% \, h^{-1}$ and $0.5 \% \, h^{-1}$. Charitos [38] reported slightly higher attrition rates of $2 \% \, h^{-1}$ for continuous operation. Such attrition rates are below the anticipated make-up rate of a calcium looping system especially if employed to capture CO_2 from cement clinker production. Furthermore, fines lost from the calcium looping system can directly be fed to the cement plant's preheater tower. Therefore, loss of sorbent by attrition does not pose a severe burden on the calcium looping process since the elutriated sorbent is reused in the clinker manufacturing process. Nonetheless, lost sorbent must be replaced by fresh sorbent make-up. In conclusion, attrition should be a subordinate issue concerning both sorbent activity (i.e. make-up) and energy related topics.

Belite formation

So far, only minor work has been conducted investigating the utilisation of cement raw meal as calcium looping sorbent. When employing cement raw meal as calcium looping sorbent an additional side reaction, the formation of silicates, can occur. Similar to sulphation, CaO is bound by the formed calcium silicates preventing the CaO from absorbing CO_2. Generally, sorbent deactivation by silicate formation can be neglected when operating with high purity limestones due to the marginal silica content. However, silica content in cement raw meal averages around $0.15 \, kg \, kg^{-1}$ or 0.33 mole of silicon per mole of CaO. Dicalcium silicate (Ca_2SiO_4), also known as belite, can be formed at temperatures above $700\,°C$ via two reaction pathways [113]. Either by reaction of CaO with SiO_2 (eq. 2.17) or directly from $CaCO_3$ releasing CO_2 (eq. 2.18). The reaction takes place at the interface between SiO_2 and CaO. Once a product layer is formed CaO is expected to be the diffusing species.

$$2\,CaO + SiO_2 \longrightarrow Ca_2SiO_4 \tag{2.17}$$

$$2\,CaCO_3 + SiO_2 \longrightarrow Ca_2SiO_4 + 2\,CO_2 \tag{2.18}$$

In 2013, Pathi et al. [134] assessed the influence of the main raw meal components (i.e. Al_2O_3, Fe_2O_3, and SiO_2) and limestone using binary mixtures of the components. Furthermore, they assessed a synthetically mixed and an industrial raw meal. In case of the binary mixture SiO_2 hardly affected the sorbent's activity, whereas the presence of Al_2O_3 or Fe_2O_3 reduced the CO_2 carrying capacity in the initial cycle significantly. This decay in sorbent activity could be attributed to a loss of surface area associated to sintering. The utilisation of heterogeneous material mixtures with comparatively large particle sizes (0.09 mm - 0.25 mm) may have hindered solid-solid reactions such as belite formation. The lack of components promoting silicate formation such as Al_2O_3 and Fe_2O_3 further impeded the belite formation. The industrial as well as the synthetic raw meal mixture showed significantly reduced CO_2 carrying capacities and formation of belite. Alonso et al. [7] investigated two natural raw meals with different Ca-Si aggregation and a synthetic raw meal generated by mixing limestone with SiO_2 nanoparticles. They found a strong initial deactivation in the first calcination carbonation cycle associated with the formation of belite. The sorbent deactivation following the initial cycle could be described by equation 2.14 assuming that only a fraction of the raw meal's CaO is available for CO_2 capture after the first cycle, whereas the remaining CaO fraction is bound as belite (Ca_2SiO_4) [7]. The fraction of lime consumed by belite formation in the first calcination depended on the Ca-Si aggregation and calcination conditions. Sorbent deactivation by belite formation increased with increasing calcination temperature or calcination time. Furthermore, raw meal with homogenous Ca-Si distribution yielded enhanced belite formation at milder calcination conditions (i.e. reduced temperature and residence time) [7]. However, at a calcination temperature of 910 °C, required for sorbent regeneration in the calcium looping process, minor differences between investigated raw meals were found for prolonged calcination times [7]. In subsequent work (in cooperation with IFK), it could be shown that raw meal with homogenous Ca-Si distribution holds a significantly increased CO_2 carrying capacity if calcined within a few seconds rather than in the range of a few minutes [12]. The results available so far highlight that the extent of belite formation and the associated sorbent deactivation depends on the residence time, Ca-Si aggregation and calcination temperature.

2.4 Reactor design

A large variety of models have been proposed to describe both the carbonator and the calciner of a calcium looping CO_2 capture system. They differ significantly in complexity ranging from simplistic 0D approaches to 1.5D models, comprising hydrodynamic calculation in combination with various reaction models, and further to full CFD simulations [125]. A comprehensive review about reactor design and modelling can be found in the work of Martinez et al. [125]. In this work, two simplistic approaches are elaborated in more detail since they offer a good trade-off between accuracy and the amount of required data or required analysis, respectively. The results regarding the active space time approaches are presented in section 6.4.6 (carbonator) and section 6.5.3 (calciner).

2.4.1   Carbonator design, active space time approach

The active space time approach developed by Charitos et al. [37, 39] represents a simplified carbonator CO_2 balance that can be used for a basic carbonator design. The gas phase
is regarded as plug flow, whereas the sorbent or solid phase is assumed to behave as in a
continuous stirred tank reactor (CSTR). Based on the operation conditions (i.e. carbonation temperature (T_{carb}), sorbent circulation rate ($\dot{N}_{CaO,loop}$), carbonator inventory (N_{CaO}),
solid residence time ($\tau_{solid,carb}$) and the utilised sorbent (i.e. sorbent reaction rate (k_{carb}) and
averaged sorbent activity ($X_{CO_2,avg}$)) the absorbed amount of CO_2 can be calculated and
consequently, the CO_2 capture efficiency. This approach has been used to describe the CO_2
capture in the carbonator with satisfactory agreement in various works focussing on the
decarbonisation of fossil fuelled power plants up to semi industrial scale [21, 22, 37, 81].
Furthermore, Arias et. al [19] applied the approach to their experiments investigating CO_2
capture from cement plants at their 30 kW electrically heated test rig obtaining satisfactory
agreement.

In steady state conditions, the amount of CO_2 removed from the gas phase must equal
the CO_2 absorbed by the CaO particles present in the carbonator. The amount of CO_2
being removed from the gas phase can be described by the product of the CO_2 capture
efficiency ($E_{CO_2,carb}$) and the molar flow of CO_2 entering the carbonator ($\dot{N}_{CO_2,carb,in}$), or the
left side of equation 2.19, respectively. Whereas, the amount of CO_2 absorbed by the sorbent
in the carbonator can be described by the reacting CaO ($N_{CaO,active}$) with its reaction rate
($\frac{dX}{dt}\big|_{carb}$). The approach assumes that the sorbent reacts with a constant reaction rate until
it reaches its CO_2 carrying capacity. The reaction rate can be expressed as a function of
the driving force of the CO_2 partial pressure towards the equilibrium CO_2 partial pressure
($\bar{y}_{CO_2} - y_{CO_2,eq}$), the average sorbent activity ($X_{CO_2,avg}$) and the apparent reaction rate (φk_{carb})
(eq. 2.21). Assuming plug flow behaviour of the gas phase, the average CO_2 concentration
of the carbonator ($\bar{y}_{CO_2}$) can be expressed by the logarithmic mean value (eq. 2.20), whereas
the equilibrium CO_2 concentration can be calculated using equation 2.2.

The reaction time required to achieve full sorbent loading is referred to as critical reaction
time (t^*_{carb}). Only particles that hold a residence time lower then the critical reaction time
have not yet reached their full CO_2 carrying capacity and are able to absorb CO_2. The active
fraction of CaO participating in CO_2 capture can be described by equation 2.22 assuming a
CSTR residence time distribution. In this equation t^*_{carb} represents the critical reaction time

and $\tau_{solid,carb}$ the residence time of the sorbent in the carbonator. Consequently, the amount of CaO absorbing CO_2 can be expressed as product of the active fraction ($f_{active,CaO}$) and the carbonator's molar CaO inventory ($N_{CaO,carb}$). Combining equation 2.19 to 2.22, the CO_2 capture efficiency can be expressed by equation 2.23.

$$\dot{N}_{CO_2,carb,in} \cdot E_{CO_2,carb} = N_{CaO,carb,active} \cdot \left.\frac{dX}{dt}\right|_{carb} \tag{2.19}$$

$$\bar{y}_{CO_2} = \frac{y_{CO_2,carb,in} - y_{CO_2,carb,out}}{\ln\left(\dfrac{y_{CO_2,carb,in}}{y_{CO_2,carb,out}}\right)} \tag{2.20}$$

$$\left.\frac{dX}{dt}\right|_{carb} = \varphi k_{carb} \cdot X_{CO_2,avg} \cdot (\bar{y}_{CO_2} - y_{CO_2,eq}) \tag{2.21}$$

$$f_{active,CaO} = 1 - exp\left(-\frac{t^*_{carb}}{\tau_{solid,carb}}\right) \tag{2.22}$$

$$E_{CO_2,carb} = \frac{N_{CaO,carb} \cdot f_{active,CaO} \cdot X_{CO_2,avg}}{\dot{N}_{CO_2,carb,in}} \cdot \varphi k_{carb} \cdot (\bar{y}_{CO_2} - y_{CO_2,eq}) \tag{2.23}$$

The fraction in equation 2.23 comprises all parameters that can be adjusted by design or plant operation (aside from sorbent selection). Hence, these are combined in one parameter, the so called active space time ($\tau_{carb,active}$), as shown in equation (2.24). To achieve a target CO_2 capture efficiency a certain active space time is required. The required amount of active space time can be obtained by adjusting (i) the sorbent circulation (i.e. active fraction of CaO in the carbonator $f_{active,CaO}$), (ii) molar carbonator inventory ($N_{CaO,carb}$) (i.e. the available reaction time of CO_2 ($\frac{N_{CaO}}{\dot{N}_{CO_2,carb,in}}$)) and/or, (iii) the sorbent activity ($X_{CO_2,avg}$) adjustable by means of sorbent make-up. The molar flow of CO_2 as well as the CO_2 concentration of the flue gas are imposed to the calcium looping CO_2 capture system by the upstream process and are therefore not adjustable. The gas solid contact factor (φ) comprises all physical resistances, such as the actual gas solid contact or diffusion resistances, occurring during carbonation [138].

$$E_{CO_2,carb} = \tau_{carb,active} \cdot \varphi k_{carb} \cdot (\bar{y}_{CO_2} - y_{CO_2,eq}) \tag{2.24}$$

2.4.2 Calciner design, active space time approach

Martinez. et al. [124] adapted the active space time approach developed by Charitos et al. [37, 39] for the carbonator to the calciner to describe its sorbent regeneration efficiency or calcination performance, respectively. The calcination or regeneration efficiency (E_{calc})

can be described by equation 2.25 as ratio of the calcined amount of $CaCO_3$ ($\dot{N}_{CaCO_3,calc,in}$ − $\dot{N}_{CaCO_3,calc,out}$) to the incoming amount of $CaCO_3$ ($\dot{N}_{CaCO_3,calc,in}$). The incoming amount of $CaCO_3$ comprises the circulating flow of sorbent exiting the carbonator as well as the make-up flow of fresh sorbent.

$$E_{calc} = \frac{\dot{N}_{CaCO_3,calc,in} - \dot{N}_{CaCO_3,calc,out}}{\dot{N}_{CaCO_3,calc,in}} \tag{2.25}$$

In steady state conditions, the calcination rate of the $CaCO_3$ present in the calciner must equal the difference of $CaCO_3$ entering and exiting the calciner. By rearranging equation 2.25, the decline of the $CaCO_3$ can be expressed by the product of the incoming molar flow of $CaCO_3$ ($\dot{N}_{CaCO_3,calc,in}$) and the calciner's calcination efficiency (E_{calc}) while the calcination rate can be expressed by the right side of equation 2.26. In this equation $N_{Ca,calc,active}$ represents the calciner's molar amount of calcium participating in the calcination reaction (i.e. calcium carbonate). The calcination reaction rate ($\frac{dX}{dt}\big|_{calc}$) can be described by equation 2.27 for elevated CO_2 partial pressures [62, 123]. TGA experiments further showed that the sorbent's calcination reaction rate remains approximately constant throughout the calcination conversion and throughout the repetitive calcination and carbonation cycles [123]. Hence, the average calcination reaction rate ($\bar{r}_{calc}$) can be obtained by integration of equation 2.27 (eq. 2.28).

$$E_{calc} \cdot \dot{N}_{CaCO_3,calc,in} = N_{Ca,calc,active} \cdot \frac{dX}{dt}\bigg|_{calc} \tag{2.26}$$

$$\frac{dX}{dt}\bigg|_{calc} = k_{calc} \cdot (1-X)^{\frac{2}{3}} \cdot (y_{CO_2,eq.} - y_{CO_2}) \tag{2.27}$$

$$\bar{r}_{calc} = \frac{k_{calc}}{3} \cdot (y_{CO_2,eq.} - y_{CO_2}) = \frac{X_{CaCO_3}}{t^*_{calc}} \tag{2.28}$$

Only a fraction of particles are participating in the calcination reaction, namely those containing $CaCO_3$. Assuming a CSTR residence time distribution of solids in the calciner, this fraction can be described by equation 2.29 using the required calcination time (t^*_{calc}) as measure for full calcination. Sorbent that holds a residence time ($\tau_{solid,calc}$) smaller than the required calcination time, are not yet fully calcined and belong to the reacting fraction of particles. The required time to achieve full calcination depends primarily on the particle's carbonate content. Hence, sorbent make-up will require longer residence times to achieve full calcination than the cycled sorbent. For low shares of sorbent make-up to circulating sorbent the difference in calcination time can be neglected [124]. Consequently, the average carbonate content ($X_{CaCO_3,avg}$) entering the calciner can be used to describe the calciner's regeneration efficiency (eq. 2.30) [124]. Moreover, for low sorbent make-up to circulating sorbent ratios the average carbonate content should match the carbonator's carbonate content fairly well.

$$f_{\text{active,CaCO}_3} = 1 - \exp\left(-\frac{t^*_{\text{calc}}}{\tau_{\text{solid,calc}}}\right) \tag{2.29}$$

$$E_{\text{calc}} = \frac{N_{\text{Ca,calc}} \cdot f_{\text{active,CaCO}_3}}{(\dot{N}_{\text{Ca,loop,in}} + \dot{N}_{\text{Ca,0,in}}) \cdot X_{\text{CaCO}_3,\text{avg}}} \cdot \frac{k_{\text{calc}}}{3} \cdot (y_{\text{CO}_2,\text{eq.}} - y_{\text{CO}_2}) \tag{2.30}$$

For given calcination conditions (i.e. calcination temperature and CO_2 concentration) the parameters that can be influenced by means of design and operation of the calciner are combined to one parameter, the calciner's active space time ($\tau_{\text{calc,active}}$).

$$E_{\text{calc}} = \tau_{\text{calc,active}} \cdot \frac{k_{\text{calc}}}{3} \cdot (y_{\text{CO}_2,\text{eq.}} - y_{\text{CO}_2}) \tag{2.31}$$

However, since the main objective of the calcium looping process is to capture CO_2 in the carbonator, the amount of circulating sorbent ($\dot{N}_{\text{Ca,loop}}$), make-up ($\dot{N}_{\text{CaCO}_3,0}$) as well as the sorbent activity will most likely be defined by the carbonator rather than the calciner. Therefore, the calciner's solid inventory ($N_{\text{Ca,calc}}$) is the main design parameter to meet the requirement of sorbent calcination once the carbonator is defined.

Chapter 3

Calcium looping CO_2 capture for clinker manufacturing

The integration options presented in this section have been developed within the *CEMCAP* project in close collaboration with the partners of the calcium looping work package. There are inherent synergies for calcium looping CO_2 capture from cement plants. Both, the cement plant and the calcium looping process rely on calcium containing solids. Hence, the deactivated calcium looping sorbent can be reutilised as feedstock for clinker manufacturing in the cement plant. Furthermore, the heat provided by the calcium looping process can be integrated into the cement plant by means of raw meal preheating or supply of power for the calcium looping's and the cement plant's auxiliary equipment. The optimal integration depends substantially on the cement plant's boundary conditions. In principle, two different calcium looping concepts can be applied to capture CO_2 from cement plants. A shallow integrated and less energy efficient option that only interacts with the clinker manufacturing process in a minor way. In this option, the calcium looping process is installed between the cement plant's raw meal mill and its preheater tower. Hence, such an option is easy to integrate or retrofit to an existing cement plant. The second option, is a deeply integrated option that is more energy efficient and interacts in a more complex way with the clinker manufacturing process. In this integration option the calcium looping process is installed between the cement plant's rotary kiln and the cement plant's preheater tower. All integration options can be characterised using four criteria - namely (i) the number of calciners or calcination steps, (ii) the type of sorbent used in the calcium looping process, (iii) the type of reactor (i.e. fluidised bed or entrained flow) employed within the calcium looping process and (iv) the amount of sorbent utilised in the calcium looping process (i.e. the integration level).

Number of calcination steps/calciners

Both the calcium looping process and the clinker manufacturing process require a calcination step. This enables the possibility to combine the calcium looping oxy-fuel calciner with the cement plant's pre-calciner. Nonetheless, operation with two calciners, an oxy-fuel fired calcium looping calciner and an air-fired pre-calciner, is also a viable option. Combining the two calcination steps to a single oxy-fuel calcination or partially shifting calcination from air to oxy-fuel will lead to significantly reduced CO$_2$ loading towards the calcium looping carbonator since CO$_2$ originating from raw meal calcination is already captured by oxy-fuel calcination. Inversely, when operating with two calciners, the share of raw meal that is calcined at air-conditions needs to be captured and calcined a second time within the calcium looping process.

Type of reactor

Two different reactor concepts are considered for calcium looping CO$_2$ capture from cement plants. The first and conventional reactor type are fluidised bed reactors due to their beneficial gas solid contact and heat transfer characteristics. Fluidised bed operation with finely milled raw meal is impractical since the finely milled raw meal belongs to the Geldard C particle classification [70]. Consequently, the raw material utilised as sorbent in the calcium looping process needs to be milled to a coarser particle fraction. Furthermore, to avoid a potential negative impact of coarser particles on the clinker quality, it is foreseen that the coarse purge from the calcium looping system is cooled down, milled to a finer particle size and reheated, whereas the fines lost through the fluidised beds cyclones can directly be fed to the cement plant. However, Thormann [170] showed that the particle size of SiO$_2$ determines the raw meal's sintering performance [100]. Hence, if a limestone-rich sorbent is used as sorbent within the calcium looping process the milling step of the calcium looping purge might be neglected and the purge can be directly fed to the cement plant's preheater tower. Such operation might especially be feasible for lower shares of sorbent purge (i.e. low integration levels) and if the fluidised bed system is operated with a small particle size in the range close to the raw meal's particle size. Direct utilisation of the purge without milling would reduce the fuel demand of the overall system.

The second and novel reactor type with respect to calcium looping are entrained flow reactors. Since entrained flow reactors operate with smaller particle sizes, the milled raw meal or a dedicated raw meal fraction can be used as sorbent in the calcium looping process and reused as feedstock for the clinker manufacturing process without further processing. Therefore, this approach might be especially suitable for deep integration configurations.

However, the gas solid contact of entrained flow is worse compared to fluidised bed reactors requiring a prolonged residence time for the sorbent to complete carbonation.

Definition of integration level

An important operation parameter that determines the operation performance and design of a calcium looping CO$_2$ capture process is the amount of make-up that is fed to the system to replace deactivated sorbent and maintain a constant, sufficiently high sorbent activity. Since the purge from the calcium looping system is used as feedstock in the cement clinker manufacturing process the make-up rate of fresh sorbent can be chosen rather freely. The term integration level (ξ_{IL}) is used to describe the degree of solid interaction between the calcium looping system and the clinker manufacturing process. It is defined by the molar amount of calcium being fed to the calcium Looping system ($\dot{N}_{Ca,CaL}$) with respect to the amount of calcium used to manufacture clinker in the cement plant's kiln ($\dot{N}_{Ca,clinker}$).

$$\xi_{IL} = \frac{\dot{N}_{Ca,CaL}}{\dot{N}_{Ca,clinker}} \tag{3.1}$$

A particular feature of calcium looping CO$_2$ capture from cement plants is that a significant share of CO$_2$ is released during calcination of carbonates present in the cement plant's raw meal. Consequently, the cement plant's CO$_2$ emissions are reduced by the share of CaCO$_3$ that is used as sorbent in the calcium looping process and fed back to the cement plant as CaO. Concurrently, this share of CO$_2$ is directly captured by oxy-fuel calcination. This circumstance can lead to very high specific make-up rates (i.e. make-up ratios) which will lead to beneficial operation conditions of the calcium looping system facilitating CO$_2$ capture. Based on the carbon intensity of the fuel burned in the cement kiln, make-up ratios up to $4\,\text{mol}\,\text{mol}^{-1}$ are possible if only CO$_2$ from the fuel combustion in the cement kiln needs to be captured in the calcium looping carbonator. That means that all CaCO$_3$ is fed to the calcium looping system, that the calcium looping purge is fully calcined and that the sensible heat of the kiln flue gas is sufficient to preheat the calcium looping purge.

Type of sorbent

Another aspect that needs to be considered when integrating the calcium looping CO$_2$ capture process into a cement plant, is the type of sorbent that is used within the calcium looping process. Depending on the cement plant's raw meal quality, sorbent from various origins are feasible. If the raw meal comprises multiple raw materials, such as limestones and marlstones, each individual raw meal fraction as well as the mixed raw meal is considered a potential calcium looping sorbent. For sure, the sorbent with the best CO$_2$ capture performance shall preferably be used respecting the framework of the particular calcium looping integration option. However, if the available stone qualities on site are considered not applicable, limestone from an external source can be employed as sorbent operating the calcium looping system at minimum make-up rates.

For the sake of simplicity two fundamentally different integration concepts will be presented in this work. However, it should be highlighted that in principle every combination of the above mentioned criteria could be feasible depending on the boundary conditions of the cement plant. In section 3.1 the so-called tail-end calcium looping integration option - a shallow integrated option using fluidised bed reactors - is presented. Whereas, in section 3.2 the so-called integrated calcium looping configuration - a deeply integrated entrained flow calcium looping configuration - is elaborated. Further process configurations are briefly presented in appendix B.

3.1   Back-end calcium looping option using fluidised bed reactors

A schematic of the tail-end or back-end calcium looping integration option using fluidised bed reactors is depicted in figure 3.1. Characteristic for the tail-end or back-end calcium looping integration is that CO_2 is captured from the clinker manufacturing process after the cement plants pre-calciner. Only minor interaction between the calcium looping CO_2 capture process and the clinker manufacturing process occurs. This integration option preferably uses limestone or a limestone rich raw meal fraction as sorbent to run the calcium looping CO_2 capture process as marlstones show lower CO_2 carrying capacities compared to limestones. Purged sorbent (i.e. CaO) is cooled down and mixed with the remaining raw meal components, proportionally replacing a fraction of the raw meal's limestone or calcium carbonate, respectively. CO_2 from initial sorbent calcination is directly captured by oxy-fuel calcination in the calcium looping calciner. Consequently, the flue gas to be decarbonised by the calcium looping carbonator consists of (i) the cement kiln's combustion flue gas, (ii) the pre-calciner's combustion flue gas as well as (iii) the CO_2 from raw meal calcination, (iv) reduced by the share of calcium carbonate being already calcined by the calcium looping process. This integration method holds an independent degree of freedom regarding the amount of sorbent, that is used in the calcium looping system (i.e. integration level). The integration level strongly influences the overall system. With increasing sorbent make-up the share of directly captured CO_2 associated with raw meal calcination

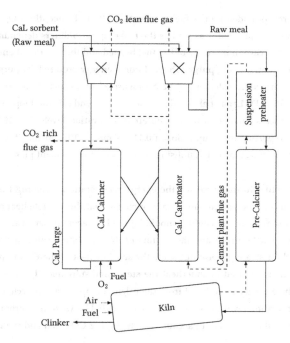

Figure 3.1: Schematic of a tail-end calcium looping CO$_2$ capture option using fluidised bed reactors. Solid streams are presented as solid lines and gas streams as dashed lines.

increases. Concurrently, the sorbent activity increases which influences the required reactor solid inventory or sorbent circulation rate between the calciner and carbonator, as well as the calciner and carbonator operation conditions. However, depending on the cement plant's boundary conditions, primarily the available raw meal composition (i.e. limestone to marlstone ratio of the cement plant's raw meal), operation with low make-up rates similar to power plant application may be favoured. The integration level indirectly determines the make-up ratio of the calcium looping system since it defines the CO$_2$ load towards the carbonator as well as the sorbent make-up flow.

Theoretically, make-up ratios up to $4\,\mathrm{mol\,mol^{-1}}$ are possible assuming that (i) all CaCO$_3$ is fed to the calcium looping system, that (ii) the calcium looping purge is exiting fully calcined, and that (iii) the sensible heat of the kiln flue gas is sufficient to preheat the calcined raw meal. However, the reheating of cooled calcium looping purge is expected to require additional energy provided by the fuel combustion [105]. Hence, the CO_2 load towards the calcium looping carbonator as well as the make-up ratio differs depending on the integration level, the carbon intensity of the fuel burned and the additional energy required to reheat the calcium looping purge. The additional energy required for purge preheating in the cement plant's pre-calciner can be compensated by means of energy recuperation from purge cooling. De Lena et al. [105] assessed the tail-end calcium looping CO_2 capture option using process simulation for five different integration levels (i.e. 15 %, 20 %, 25 %, 50 % and 80 %) reporting make-up ratios of $0.11 \, \text{mol} \, \text{mol}^{-1}$, $0.16 \, \text{mol} \, \text{mol}^{-1}$, $0.21 \, \text{mol} \, \text{mol}^{-1}$, $0.60 \, \text{mol} \, \text{mol}^{-1}$ and $1.84 \, \text{mol} \, \text{mol}^{-1}$, utilising coal as fuel in the cement plant.

De Lena et al. [105] further assessed the energy recuperation including the steam cycle design by means of process simulation and concluded that the sensible heat of the carbonator flue gas and from purge cooling should be used as 'economiser heat'. Furthermore, the carbonation reaction heat and the cooling of the circulating solids in the carbonator should be used for steam generation, while the sensible heat of the CO_2 rich calciner flue gas should initially be used to superheat the steam and subsequently to 'economise' the feed water. The electricity produced from the calcium looping steam cycle increases with decreasing integration level as more solids are circulated between the carbonator and calciner (i.e. reheated and cooled) to account for the higher CO_2 load and the lower sorbent activity [105].

3.2   Integrated calcium looping option using entrained flow reactors

A schematic of the integrated calcium looping option using entrained flow reactors is presented in figure 3.2. Characteristic for the deep integration is that CO_2 from raw meal calcination is almost completely captured by oxy-fuel calcination. Only a minor fraction of the raw meal, that is required to cool down the kiln flue gas, is not calcined in oxy-fuel environment. The utilisation of entrained flow reactors enables direct feeding of the calcium looping purge into the cement plant's rotary kiln.

In this integrated calcium looping option, the CO_2 from the kiln flue gas is captured in the calcium looping carbonator, whereas CO_2 from raw meal calcination is directly captured by oxy-fuel calcination. The kiln flue gas needs to be cooled down from approx. 1450 °C to carbonation temperatures of around 650 °C. Since the kiln flue gas features high tempera-

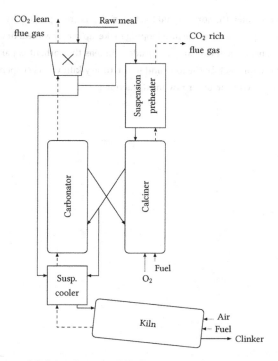

Figure 3.2: Integrated Calcium Looping CO_2 Capture option using entrained flow reactors. Solid streams are presented as solid lines and gas streams as dashed lines.

tures in combination with high sulphur and alkali content as well as high particle loading, a dedicated partial flow of raw meal is used to cool down the flue gas in a suspension heat exchanger. Ideally, a raw meal fraction with a poor CO_2 carrying capacity is used to cool down the kiln flue gas to enhance the activity of the sorbent in the carbonator. The cooled kiln flue gas is fed into the entrained flow carbonator where it is mixed with regenerated sorbent and decarbonised by carbonation of free CaO of the sorbent. The poorer gas solid contact of entrained flow reactors requires longer residence times. The required contact times could be achieved by a goose neck shaped carbonator design or to a limited extent by enhanced sorbent loading increasing the amount of CaO available for carbonation in the entrained flow carbonator. After the carbonator, a cyclone separates the loaded sorbent from the CO_2 depleted flue gas. The sorbent is directed into the oxy-fuel calciner, whereas the decarbonised flue gas is used as carrier gas in the raw meal mill. After regeneration in the oxy-fuel fired calciner, the sorbent and the CO_2 rich gas are separated by a cyclone. The CO_2 rich gas stream preheats the calcium looping make-up (i.e. raw meal or raw meal component) in a suspension heat exchanger cascade, whereas the sorbent is partially directed towards the carbonator to close the loop and to the rotary kiln via its suspension cooler to intensify the mixing with the other raw meal fractions.

Chapter 4

Methodology

In this chapter the experimental set-up and methodology of the conducted screening and characterisation of various potential calcium looping sorbents, the fluidised bed experiments and entrained flow experiments including their respective, subsequent sorbent analyses using thermogravimetry (TGA) are presented.

4.1 Sorbent characterisation and sorbent screening

Limestone from various origins as well as different raw meal mixtures representing raw meals utilised in Europe and their respective raw materials were investigated regarding their cyclic CO_2 carrying capacity by thermogravimetric analysis. Two size cuts of the Lhoist limestone from Messinghausen with a nominal particle size of 100 μm to 300 μm (RK13) and 300 μm to 700 μm (RK37) were assessed as well as three size cuts of the limestone fraction of the Vernasca raw meal ranging from 0 μm to 100 μm (VC01), from 100 μm to 420 μm (VC14) and from 420 μm to 1000 μm (VC410). The chemical composition of the investigated sorbents are summarised in table 4.1 (limestone based) and table 4.2 (raw meal based). The particle size of the limestone based sorbents and the mixed raw meals are presented in figure 4.1 and 4.2, respectively. A comprehensive overview of the utilised sorbents including the raw meal components is given in the appendix in section A.2.

Four different TGA programmes were used to examine how the sorbent reactivity is affected by the calcination and carbonation conditions. Their characteristic variation in operation conditions are summarised in table 4.3. In the reference run, representing dry air calcination, the sorbent was calcined in pure nitrogen at 850 °C for 10 min and carbonated for 10 min with a CO_2-N_2 mixture containing $0.15\,\mathrm{m}^3\,\mathrm{m}^{-3}$ CO_2. Compared to the reference case $0.15\,\mathrm{m}^3\,\mathrm{m}^{-3}$ of nitrogen was replaced by steam throughout the analysis in the H_2O run.

Table 4.1: Chemical composition of utilised limestone based sorbents.

	x_{CaO}	x_{SiO_2}	$x_{Fe_2O_3}$	$x_{Al_2O_3}$	x_{MgO}	x_{CO_2}	x_{Others}
				$kg\,kg^{-1}$, wf			
RK	0.5465	0.0071	0.0006	0.0012	0.0079	0.4357	0.0009
Verdal	0.5337	0.0540	0.0006	0.0019	0.0034	0.4042	0.0023
EnBW	0.5584	0.0030	0.0008	0.0013	0.0026	0.4337	0.0003
Saabar	0.5222	0.0399	0.0018	0.0025	0.0000	0.4243	0.0093

Table 4.2: Chemical composition of utilised raw meal based sorbents.

	x_{CaO}	x_{SiO_2}	$x_{Fe_2O_3}$	$x_{Al_2O_3}$	x_{MgO}	x_{CO_2}	x_{Others}
				$kg\,kg^{-1}$, wf			
Vernasca raw meal	0.4127	0.1673	0.0233	0.0356	0.0121	0.3307	0.0184
Vernasca limestone	0.5035	0.0298	0.0047	0.0091	0.0176	0.4315	0.0039
Vernasca marl	0.3732	0.2245	0.0165	0.0436	0.0120	0.3024	0.0278
Bilbao raw meal	0.4007	0.1439	0.0195	0.0628	0.0138	0.3349	0.0245
Rosenheim raw meal	0.3882	0.1402	0.0198	0.0564	0.0450	0.3300	0.0204
Geseke raw meal	0.4345	0.1455	0.0089	0.0227	0.0044	0.3704	0.0135
Geseke North	0.4408	0.1395	0.0077	0.0238	0.0045	0.3705	0.0132
Geseke South	0.4352	0.1427	0.0072	0.0242	0.0044	0.3741	0.0122
Geseke Corrective	0.4062	0.1914	0.0081	0.0273	0.0050	0.3464	0.0156
Rumelange raw meal	0.4576	0.1367	0.0253	0.0272	0.0018	0.3371	0.0142
Rumelange marl Rouge	0.4194	0.1768	0.0408	0.0189	0.0007	0.3341	0.0093
Rumelange marl Grise	0.4390	0.1612	0.0227	0.0180	0.0005	0.3498	0.0089
Rumelange limestone	0.4941	0.0847	0.0093	0.0098	0.0006	0.3943	0.0072
Rumelange limestone HP	0.4177	0.1914	0.0081	0.0273	0.0050	0.3464	0.0041
Rumelange Catalyst	0.0010	0.4457	0.0080	0.5150	0.0000	0.0019	0.0284

In the oxy-fuel programme, sorbent was calcined at 920 °C in a CO_2-steam mixture containing $0.85\,m^3\,m^{-3}$ CO_2 and $0.15\,m^3\,m^{-3}$ H_2O for 10 min and carbonated with a CO_2-steam-N_2 mixture containing $0.15\,m^3\,m^{-3}$ CO_2, $0.15\,m^3\,m^{-3}$ H_2O and $0.70\,m^3\,m^{-3}$ N_2 for 10 min. In the fourth routine, the reference programme was extended with an initial segment in which the sample is heated up and held in pure CO_2 at 850 °C for 20 min to allow potential belite formation (direct pathway) to be completed. Subsequently, the sample is calcined for 10 min in pure N_2 followed by the reference programme. For all experiments the TGA's heating rate was set to $200\,K\,min^{-1}$, and a cooling rate of approx. $50\,K\,min^{-1}$ was obtained in the range of 920 °C to 600 °C.

Figure 4.3 illustrates exemplarily the determination of the CO_2 carrying capacity for three different cycles. The CO_2 carrying capacity was evaluated as subsequently described. The

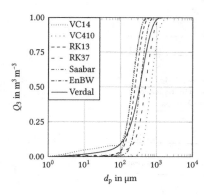
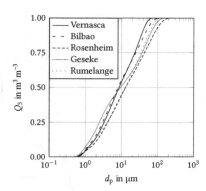

Figure 4.1: Cumulative particle size distribution (Q_3) vs. particle diameter (d_p) for the investigated limestone based sorbents.

Figure 4.2: Cumulative particle size distribution (Q_3) vs. particle diameter (d_p) for the investigated raw meals.

CaO content of the sample was determined by the mass loss during the first calcination assuming that the mass loss is solely attributed to the release of CO_2 from $CaCO_3$ calcination. This procedure is justifiable by the small amount of magnesium present in the investigated sorbents. The CO_2 uptake was determined at the end of the carbonation phase rather than at the intersection of the fast chemical controlled regime and diffusion controlled regime. This approach was chosen since the cyclic TGA experiments were operated with a sample mass of 12 mg to obtain a better signal to noise ratio and to ensure adequate resolution at lower sorbent activities. However, this comes with the drawback that the transition from the chemical controlled regime to the diffusion controlled regime is less distinct. Nonetheless, as can be seen from figure 4.3 this imposes only a minor inaccuracy on the system as the difference in the determined CO_2 carrying capacity is rather low. The results concerning the sorbent screening for potential calcium looping utilisation are presented in chapter 5.

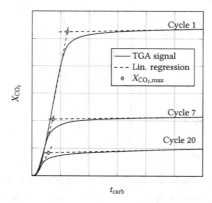

Figure 4.3: Determination of CO_2 uptake (X_{CO_2}) for various cycles (dry air calcination).

Table 4.3: TGA routines utilised for sorbent screening and characterisation.

Analysis (acronym)	Reaction atmosphere	
	Calcination	Carbonation
dry air calcination (Air)	$T_{calc} = 850\,^\circ C$ $y_{CO_2} = 0.00\,\mathrm{m^3\,m^{-3}}$ $y_{N_2} = 1.00\,\mathrm{m^3\,m^{-3}}$ $y_{H_2O} = 0.00\,\mathrm{m^3\,m^{-3}}$	$T_{carb} = 650\,^\circ C$ $y_{CO_2} = 0.15\,\mathrm{m^3\,m^{-3}}$ $y_{N_2} = 0.85\,\mathrm{m^3\,m^{-3}}$ $y_{H_2O} = 0.00\,\mathrm{m^3\,m^{-3}}$
humid air calcination (H$_2$O)	$T_{calc} = 850\,^\circ C$ $y_{CO_2} = 0.00\,\mathrm{m^3\,m^{-3}}$ $y_{N_2} = 0.85\,\mathrm{m^3\,m^{-3}}$ $y_{H_2O} = 0.15\,\mathrm{m^3\,m^{-3}}$	$T_{carb} = 650\,^\circ C$ $y_{CO_2} = 0.15\,\mathrm{m^3\,m^{-3}}$ $y_{N_2} = 0.70\,\mathrm{m^3\,m^{-3}}$ $y_{H_2O} = 0.15\,\mathrm{m^3\,m^{-3}}$
Oxy-fuel calcination (Oxy)	$T_{calc} = 920\,^\circ C$ $y_{CO_2} = 0.85\,\mathrm{m^3\,m^{-3}}$ $y_{N_2} = 0.00\,\mathrm{m^3\,m^{-3}}$ $y_{H_2O} = 0.15\,\mathrm{m^3\,m^{-3}}$	$T_{carb} = 650\,^\circ C$ $y_{CO_2} = 0.15\,\mathrm{m^3\,m^{-3}}$ $y_{N_2} = 0.70\,\mathrm{m^3\,m^{-3}}$ $y_{H_2O} = 0.15\,\mathrm{m^3\,m^{-3}}$
CO$_2$ pre-treatment (CO$_2$)	$T_{calc} = 850\,^\circ C$ $y_{CO_2} = 0.00\,\mathrm{m^3\,m^{-3}}$ $y_{N_2} = 1.00\,\mathrm{m^3\,m^{-3}}$ $y_{H_2O} = 0.00\,\mathrm{m^3\,m^{-3}}$	$T_{carb} = 650\,^\circ C$ $y_{CO_2} = 0.15\,\mathrm{m^3\,m^{-3}}$ $y_{N_2} = 0.85\,\mathrm{m^3\,m^{-3}}$ $y_{H_2O} = 0.00\,\mathrm{m^3\,m^{-3}}$

4.2   Fluidised bed calcium looping CO_2 capture

Various integration levels between the cement plant and a tail-end fluidised bed calcium looping system have been assessed using University of Stuttgart's 200 kW pilot plant. A high purity limestone from western Germany with a nominal particle size between 100 μm and 300 μm (RK13) was used as sorbent. Two different Columbian hard coals were used as fuel. Initially the El Cerrejon coal was used for the experiment. Since the El Cerrejon coal was no longer dispatchable, the La Loma coal was employed for subsequent experiments. The La Loma mine is located in close proximity to the El Cerrejon mine, around 250 km south west. The two coals differ mainly in terms of sulphur content. Their respective ultimate analyses are presented in table 4.4, whereas the sorbent chemical composition can be obtained from table 4.1.

Table 4.4: Chemical composition of coal utilised for the fluidised bed experiments.

	Y_C	Y_H	Y_O	Y_N	Y_S	Y_{ash}	Y_{H_2O}
	kg kg^{-1}, waf					kg kg^{-1}, wf	kg kg^{-1}, ad*
El Cerrejon	0.803	0.049	0.123	0.019	0.006	0.096	0.074
La Loma	0.776	0.053	0.144	0.016	0.011	0.091	0.074

*ad: air dried as used in experiments

4.2.1   Fluidised bed pilot facility (MAGNUS)

University of Stuttgart's 200 kW fluidised bed pilot plant located at the Institute of Combustion and Power Plant Technology consists of three refractory lined fluidised bed reactors that can be interconnected variously resulting in a CFB-CFB or a BFB-CFB configuration. A schematic of the CFB-CFB configuration is presented in figure 4.4, whereas figure 4.5 depicts the BFB-CFB arrangement. In both configurations, the calcination reactor is a circulating fluidised bed reactor while depending on the configuration either the CFB carbonator or the BFB carbonator is deployed. Both CFB reactors are 10.8 m high and have an inner diameter of approx. 200 mm. The calciner's inner diameter increases gradually from 120 mm at the reactor bottom to 200 mm at a reactor height of 3.250 m reaching its final inner diameter of 210 mm at 4.85 m. The carbonator's initial inner diameter of 330 mm is reduced at approx. 1 m to 220 mm. This reactor design allows increased residence times in the bottom section of the reactor or its dense bed region. The BFB carbonator has a height of 6 m and an inner diameter of 330 mm. In a transition zone at the bottom of the reactor, the inner diameter is eccentrically reduced from 300 mm to 120 mm over a height of 350 mm. Subsequent to

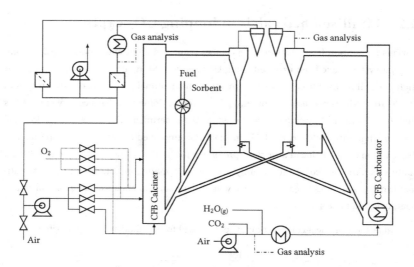

Figure 4.4: CFB-CFB configuration of University of Stuttgart's 200 kW calcium looping pilot plant at the Institute of Combustion and Power Plant Technology (IFK).

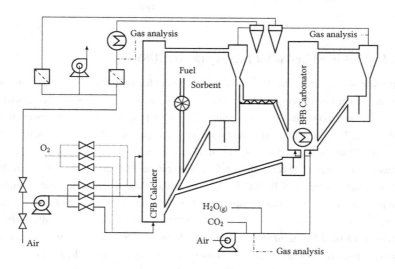

Figure 4.5: BFB-CFB configuration of University of Stuttgart's 200 kW calcium looping pilot plant at the Institute of Combustion and Power Plant Technology (IFK).

each fluidised bed riser a cyclone separates the entrained solids from the gas. Additionally, a protective cyclone is installed before each bag filter to separate finer particles and to
protect the bag filter in case of malfunction. Solids separated by the cyclones are directed
back to the reactors via loop seals, whereas the solids separated by the protective cyclone
are collected in a bin. For each reactor, temperature and pressure is measured along the
reactor height, the solid circulation lines and exhaust gas lines by thermocouples (type N)
and pressure transducers (piezoresistive). In case of CFB-CFB configuration, solids separated by the cyclone are either recycled internally or directed to the other reactor. The
respective share of external and internal circulation is controlled via a cone valve in the
reactor's loop seal. This design allows an independent investigation of the circulation rate
from solid entrainment. For the BFB-CFB configuration, solids are fed from the calciner
to the carbonator via a screw feeder and cycled back to the calciner via a loop seal at the
bottom of the carbonator. Solid circulation is solely controlled by the screw feeder's rotational speed. For calcium looping operation, fuel and limestone are continuously fed into
the calciner by means of gravimetrically controlled screw feeders and a rotary valve. The
carbonator is fluidised with synthetic flue gas generated by mixing CO_2, air and steam. For
operation of the oxy-fuel calciner, CO_2 rich exhaust gas is recycled and mixed with oxygen
for fuel combustion. The gas mixture can be fed at three stages to control the combustion
performance in the calciner. At each stage the flow of recirculated flue gas as well as its
oxygen feed can be adjusted individually. Volume flows, gas composition at the calciner's
recirculation line as well as at the carbonator inlet and outlet, pressure and temperature, and
parameters derived from these values are continuously monitored, calculated and recorded,
whereas solid samples are collected periodically. The circulation rate between the reactors
is constantly monitored by microwave sensors (CFB-CFB configuration) or the conveyor
screw's rotational speed (BFB-CFB configuration) and manually measured and verified by
means of solid accumulation in a dedicated measuring section. The reactors' solid inven-
tories are calculated based on the pressure drop over each reactor part with its respective
cross-section and summed up. Subsequently, the molar calcium inventory was determined
based on the chemical analysis of the collected bed samples. The reference temperatures of
the reactors were calculated by integrating and averaging the temperature measurements
along the reactor height, since the thermocouples are not installed equidistantly along the
reactor height. As the oxidation gas is fed to the calciner in a staged manner, the integration starts after all oxidant is injected into the calciner (i.e. after the third stage or at 4 m,
respectively). For the CFB carbonator the temperature is integrated from the gas distributor
to the reactor outlet, whereas for the BFB carbonator the temperature is integrated along

the carbonator bed. The gas composition is measured continuously by non-dispersive in-frared spectroscopy (CO, CO_2, SO_2, NO_x), paramagnetism (O_2) and impact jet psychrometry (H_2O).

Solid Sampling

Solid samples are collected from (i) the loop seals, (ii) the secondary/protective cyclones and (iii) from the bag filter. Solids separated from the secondary cyclone as well as the bag filter are collected in a bin. To ensure the collection of a representative sample, the bin of the secondary cycle was exchanged before and after each experiment. This procedure allows also to assess the loss of solids by weighing the mass accumulated during the experimental period (eq. 4.1). Since the bag filter accumulated marginal amount of solids this method was not applied to the bag filter.

$$\dot{M}_{loss} = \frac{dM}{dt} \approx \frac{\Delta M}{\Delta t} \tag{4.1}$$

Solid samples of the circulating sorbent were collected at the loop seal of the respective reactor. A schematic of the solid sampling from the loop seals is depicted in figure 4.6. The solid sample is collected from the standpipe side (i.e. left side in the schematic). Initially, the ball valve is opened until fresh, hot bed material is exiting through the sampling pipe and closed again. The actual sample cools down in the sampling pipe, sealed by the ball valve towards the ambient air and by a fraction of particles towards the circulating bed material. The sample volume is chosen to obtain sufficient distance between the loop seal bottom and the actual sample material to avoid potential back mixing with circulating solids. After approx. 20 min the cooled sample can be poured in a sample vessel. The sealing particle fraction remaining in the sampling pipe is discharged during the next sampling step. Likewise, solids required to determine the bulk density of the circulating bed material are collected.

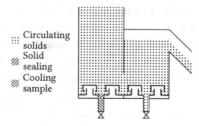

Figure 4.6: Schematic of circulating solid sample collection.

Gas Analysis

The gas compositions were continuously measured at (i) the calciner's flue gas duct (after the gas cooler), (ii) in the carbonator's synthetically mixed flue gas feed (before steam injection) and (iii) in the carbonator's exhaust gas duct. CO, CO_2, SO_2 and NO_x were measured by non-dispersive infrared spectroscopy, O_2 by paramagnetism and H_2O by impact jet psychrometry. Additionally, a zirconia probe was used to measure the calciner's and BFB carbonator's outlet oxygen concentration. Gas sampling was conducted according to DIN EN 14792 [71] using a heated candle filter to separate particles from the sample gas as well as a heated PTFE hose and heated pump to avoid condensation before the actual gas cooler. The particle filters were periodically cleansed by flushing nitrogen countercurrently. The calibration of the gas analysers was checked daily throughout the experimental campaigns and recalibrated if necessary. Table 4.5 summarises the measured species for the respective measurement location, whereas figure 4.7 presents the set-up of the gas sampling line.

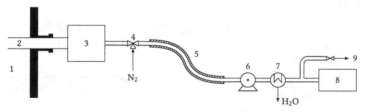

1) Flue gas duct   2) Sampling probe 3) Heated filter   4) N_2 backflushing 5) Heated PTFE hose
6) Heated pumpe 7) Condensor   8) Gas analyser   9) Bypass

Figure 4.7: Schematic of the gas sampling line following DIN EN 14792 [71].

Table 4.5: Measured gas species and their measuring principle.

Reactor	Location	Species	Measurement principle
Calciner	outlet	CO_2, CO, NO_X, SO_2	NDIR
		O_2	paramagnetism
		H_2O	impact jet psychrometry
Carbonator	inlet	CO_2, CO, O_2	NDIR
		O_2	paramagnetism
	outlet	CO_2, CO, O_2	NDIR
		O_2	paramagnetism

Solid flow measurement

The circulation rate between the reactors was measured continuously by microwave sensors measuring the particles' reflection intensity and frequency. While these sensors are sufficient to reliably operate the pilot plant, their measurement is not accurate, since they can only be calibrated at ambient conditions. As a consequence, the solid circulation between the reactors was determined periodically by accumulation of the solids in a dedicated measurement volume. Based on the time required to fill (t_{accu}) a dedicated measurement volume (V_{accu}), and the bulk density of the circulating solids ($\rho_{solid,bulk}$) the circulating mass flow between the reactors was determined using equation 4.2. Schematics of the measurement sections are depicted in figure 4.8 and figure 4.9. Figure 4.8 shows the CFB reactor's internal and external measurement section, while figure 4.9 presents the measurement section of the screw conveyor connecting the CFB calciner with the BFB carbonator. Through a peephole next to the overflow the accumulation of the measurement section can be monitored and timed.

$$\dot{M}_{loop} = \frac{V_{accu}}{t_{accu}} \cdot \rho_{solid,bulk} \tag{4.2}$$

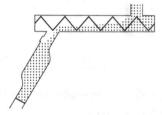
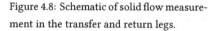

Figure 4.8: Schematic of solid flow measurement in the transfer and return legs.

Figure 4.9: Schematic of solid flow measurement using the cooled conveyor screw.

Volume flow measurement

Volume flows in the exhaust gas ducts of each reactor, the calciner's main gas feed duct as well as the secondary and tertiary stage are measured by impeller anemometers. The calciner's primary duct gas flow is calculated by subtracting the secondary and tertiary gas from the main duct flow. Other auxiliary volume flows such as loop seal fluidisation, oxygen supply to the three gas feeding stages of the calciner and quench air cooling down the carbonator's flue gas were controlled using mass flow controllers. Steam fed to the synthetic flue gas was measured using a vortex flow sensor.

4.2.2 Analysis of sorbent samples collected within the fluidised bed calcium looping experiments

Solid samples collected during the fluidised bed calcium looping experiments have been evaluated by TGA analysis regarding their CO_2 carrying capacities and calcium carbonate content using an inductively heated Linseis PT 1600 thermogravimetric analyser. An exemplary run of a TGA analysis is depicted in figure 4.10. After the sample was weighed in, it was dehydrated at 480 °C for 30 min and subsequently calcined at 850 °C for 20 min. Both, dehydration and calcination was carried out in pure nitrogen. The calcined sample was then carbonated at 650 °C for 20 min with CO_2/N_2 mixture containing $0.13\,m^3\,m^{-3}\,CO_2$ and $0.87\,m^3\,m^{-3}\,N_2$. The total volume flow of the reaction gases were $6\,L\,h^{-1}$. The severe changes in mass between the various analysis steps result from the electromagnetic force imposed on the weighting system by the inductive heating system. Hence, reference sections have been applied before and after each analysis step at a temperature of 200 °C for 20 min to determine the sorbent's mass loss or gain between two reference steps. However, mass gain due to CO_2 capture in the carbonation step can immediately be determined since the inductive force stays constant for a constant temperature. The molar CO_2 uptake during carbonation can be calculated by the mass difference between the current mass ($M(t_{carb})$) and the reference mass before the gas atmosphere is switched to carbonating conditions ($M(t_{Ref,carb})$) using equation 4.3. A comparison between the mass gain determined during carbonation (i.e. 650 °C) and the mass gain calculated using the two reference steps (i.e. 200 °C) yields good agreements indicating the validity of the chosen analysis method.

For the evaluation of the sample, it was assumed that the sample composes solely of calcium hydroxide ($Ca(OH)_2$), calcium carbonate ($CaCO_3$) and calcium oxide (CaO). Hence, the molar amount of calcium or calcium oxide can be calculated by equation 4.4. The nominator in equation 4.4 represents the mass of the fully calcined and dehydrated sample. The evolution of the sorbent's CO_2 carrying capacity during carbonation can be described by equation 4.5. The sorbent's CO_2 carrying capacity ($X_{CO_2,avg}$) was determined by the intersection of the kinetically controlled reaction regime and the diffusion controlled reaction regime (figure 4.11). Both linear functions have been determined by linear regression. The carbonation reaction in the chemical controlled reaction regime has been determined by linear regression around the infliction point of the mass signal or its derivative's maximum, respectively. Whereas, the last 2 min of the carbonation phase have been used for the linear regression of the diffusion controlled regime.

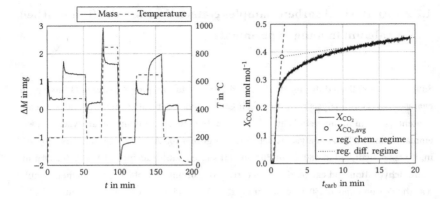

Figure 4.10: Exemplary sorbent analysis for samples collected during fluidised bed calcium looping experiments.

Figure 4.11: Determination of carbonation CO_2 carrying capacity.

$$N_{CO_2}(t_{carb}) = \frac{M(t_{carb}) - M(t_{Ref,carb})}{\tilde{M}_{CO_2}} \qquad (4.3)$$

$$N_{CaO,sample} = \frac{M_{0,sample} - \Delta M_{deyhd} - \Delta M_{calc}}{\tilde{M}_{CaO}} \qquad (4.4)$$

$$X_{CO_2}(t_{carb}) = \frac{N_{CO_2}(t_{carb})}{N_{CaO,sample}} \qquad (4.5)$$

4.3 Entrained flow calcium looping CO_2 capture

The entrained flow calcination experiments presented in this work were conducted at the so called DIVA facility at the Institute of Combustion and Power Plant Technology. Both, air calcination and oxy-fuel calcination has been investigated using four European raw meal qualities (Vernasca, Bilbao, Rumelange, and Geseke) as well as a high purity limestone fraction belonging to the Vernasca raw meal. The assessed operation conditions are summarised in table 4.6. The sorbent's composition can be obtained from table 4.2. Their respective particle size distribution can be obtained from figure 4.2 and are presented in more detail in appendix A. Subsequent to the entrained flow experiments the collected sorbent samples were analysed regarding their CO_2 carrying potential as well as their calcination degree. The effect of CO_2 partial pressure and carbonation temperature has been assessed

Table 4.6: Experimental conditions of entrained flow calcination experiments.

Calcination	T_{calc}	y_{CO_2}	y_{O_2}	Carrier gas
Air	860 °C	0.26 m^3 m^{-3}	0.03 m^3 m^{-3}	N$_2$
Oxy-fuel	900 °C 920 °C	0.90 m^3 m^{-3}	0.03 m^3 m^{-3}	CO$_2$

for two raw meals at a reference calciner residence time (i.e. τ_{calc} ≈ 4 s). The two raw meals represent a heterogeneous raw meal mixture (i.e. high purity limestone and marlstone) as well as a marl type raw meal. The results concerning entrained flow calcium looping are presented in chapter 7.

4.3.1 Entrained flow reactor (DIVA)

The entrained flow reactor used to assess the sorbent performance of raw meal based sorbents for an entrained flow calcium looping system is presented in figure 4.12. The facility is 12.4 m in height and has an inner diameter of 70.3 mm. The reactor is electrically heated by 17 pairs of resistance heatings which can be controlled individually. Additionally, natural gas combustion can be employed to increase the heat release inside the reactor. The temperature along the reactor height is measured using type N thermocouples, whereas the pressure is recorded using piezoresistive pressure transducers. Raw meal or sorbent feeding is realised by a self-constructed feeding system comprising of a weight controlled screw feeder and a venturi nozzle. Raw meal is dosed into the venturi nozzle behind its constriction to be dispersed by a carrier gas stream and fed to the bottom of the entrained flow reactor using a conductive hose to avoid solid accumulation due to static charging of the hose. Depending on the investigated reaction atmosphere either N$_2$ (air calcination) or CO$_2$ (oxy-fuel calcination) is used as carrier gas in order to avoid limitations of the calcination atmosphere by the carrier gas. Subsequent to the reactor, gas and solids are quenched to below 500 °C using compressed air in order to avoid any recarbonation of the sorbent as well as further calcination. The solids are then separated from the gas by a cyclone and collected in a sample vessel. Sorbent accumulation at the cyclone wall is avoided by constant vibration. A potential back flow of the purge gas towards the reactor is monitored by the pressure difference of the exit duct and adjusted by means of an ID fan providing sufficient negative pressure at the cyclone outlet. Before quenching, sample gas is extracted and fed to a NDIR gas analyser. Gas sampling is conducted according to DIN 14792 [71] similar to the fluidised bed experiments. Air calcination experiments were conducted at a reaction

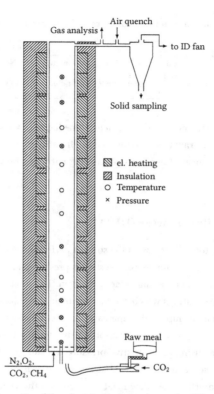

Figure 4.12: Schematic of the electrically heated entrained flow reactor DIVA.

temperature of 860 °C, representing common pre-calciner temperatures in a cement plant, with a CO_2 concentration of approx. $0.26\,m^3\,m^{-3}$, representing the logarithmic average CO_2 concentration of a cement plant's pre-calciner. Oxy-fuel experiments were operated with a CO_2 concentration of approx. $0.90\,m^3\,m^{-3}$ at two reaction temperatures (900 °C and 920 °C). All experiments were conducted with a constant excess oxygen concentration of approx. $0.03\,m^3\,m^{-3}$. A reference natural gas feed of $0.2\,m^3\,h^{-1}$ was used for the characterisation of the various raw meals at different operation conditions such as residence time, reaction atmosphere and calcination temperature. Additionally, experiments investigating the influence of the fuel feeding rate (i.e fuel combustion) on the sorbent's calcium looping properties have been conducted.

4.3.2 Conduction of entrained flow calcination experiments

Prior to raw meal feeding the reaction atmosphere and superficial gas velocity is adjusted by means of volume flows being fed to the reactor. The calcination temperature is set and controlled by the resistance heatings. Once raw meal feeding is started an immediate raise in CO_2 concentration can be observed which is associated with sorbent calcination. The system requires a few minutes to equalise with respect to temperature, before solid samples can be collected. After sample collection the raw meal feed is stopped and the reactor is flushed with nitrogen to remove particles that have potentially accumulated at the reactor bottom due to poor entrainment. This has occurred to some extent at low superficial velocities. Also, the gas sample line's particle filter is flushed with nitrogen to avoid sorbent accumulation and potential carbonation falsifying the gas measurement. A gradual decrease in CO_2 concentration towards the initial CO_2 concentration can be observed once the raw meal feed has been stopped. This decrease can be used to monitor the quality of the conducted experiments. However, in case of oxy-fuel operation, deviation from the

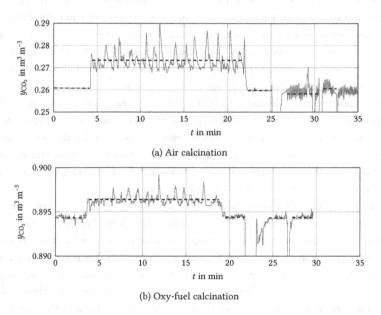

(a) Air calcination

(b) Oxy-fuel calcination

Figure 4.13: Time trend of entrained flow calcination operation. CO_2 concentration (y_{CO_2}) vs. experiment time (t) for (a) air calcination, (b) oxy-fuel calcination.

initial CO_2 concentration could be observed due to a limited resolution in the upper end of the measurement range (accuracy $0.01\,m^3\,m^{-3}$) and a comparatively small increment of CO_2 originating from sorbent calcination. The share of CO_2 emerging from raw meal calcination on the overall CO_2 concentration is especially low for high superficial velocities as the overall flow of CO_2 increases. An exemplary time trend of the experiments is depicted in figure 4.13a (air calcination) and in figure 4.13b (oxy-fuel calcination). The average CO_2 concentration for the respective steps of the experiments are depicted as dashed lines.

4.3.3   Analysis of sorbent samples collected within the entrained flow experiments

The entrained flow samples were analysed in a similar way to the fluidised bed samples. However, a new customised, radiative heated thermogravimetric analyser (Linseis PT 1100) was employed for the analysis of the entrained flow calcined solid samples. Since the samples were only partially calcined and as longer residence times at elevated temperatures may lead to silicate formation of raw meal samples, the samples were initially carbonated and subsequently calcined. By default, the carbonation was carried out at $650\,°C$ for $20\,min$ with a CO_2 concentration of $0.1\,m^3\,m^{-3}$ and a steam concentration of $0.1\,m^3\,m^{-3}$ in nitrogen. Calcination was performed at $860\,°C$ in pure N_2 for $10\,min$ to ensure full calcination. The total sample volume flow was $100\,mL\,min^{-1}$ with a protective gas flow of $50\,mL\,min^{-1}$. The reference programme is depicted in figure 4.14. The TGA analyses were conducted with a sample mass of approx. $6.5\,mg$ to improve the signal to noise ratio. As a consequence, the transition from fast reaction regime to diffusion controlled regime is less distinct. However, only minor deviations of the CO_2 carrying capacity determined at the end of the carbonation phase or at the intersection of the fast reaction and the diffusion regime and the end of the carbonation phase have been observed (figure 4.14).

For the investigation of the influence of the CO_2 partial pressure and the carbonation temperature on the carbonation reaction of the entrained flow calcined raw meal the reference analysis has been slightly altered. The sample was heated in nitrogen atmosphere to the carbonation temperature ($600\,°C$, $650\,°C$, $700\,°C$). Subsequently, the reaction atmosphere was switched to carbonation conditions with $0.10\,m^3\,m^{-3}$ steam and the respective CO_2 concentration (i.e. $0.05\,m^3\,m^{-3}$, $0.10\,m^3\,m^{-3}$, $0.20\,m^3\,m^{-3}$) in nitrogen. Since the sample is heated in pure nitrogen to its carbonation temperature the sample starts calcining during the heat up (figure 4.15). The generated amount of CaO needs to be considered when evaluating the TGA analysis. Furthermore, experiments with a reduced sample mass of approx.

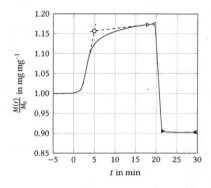
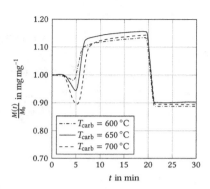

Figure 4.14: Analysis routine for the determination of characteristic recarbonation values of the entrained flow calcined sorbent samples.

Figure 4.15: Analysis routine for the assessment of the influence of CO_2 partial pressure and carbonation temperature of the entrained flow calcined raw meal samples.

2 mg allowing an assessment of the influence of CO_2 partial pressure and temperature on the carbonation kinetics have been conducted. In the framework of the kinetic assessment a fourth CO_2 concentration ($0.30\,\mathrm{m^3\,m^{-3}}$) was investigated.

Hereafter, the evaluation of the TGA measurements analysing the entrained flow calcined sorbent samples ist presented. The CO_2 content ($x_{CO_2,EF}$) of the entrained flow calcined raw meal sample (Index: EF) is determined by the mass loss associated with calcination (ΔM_{calc}) using equation 4.6. The term $M_{0,sample}$ refers to the initial weight of the sample. Based on the CO_2 content of the sample ($x_{CO_2,EF}$), its CaO content before carbonation ($x_{CaO,EF}$) is calculated based on the raw meal's chemical analysis (Index: raw meal) using equation 4.7. The CO_2 uptake during carbonation is determined as fraction of the mass gain of the sample (ΔM_{carb}) and the sample's initial mass ($M_{0,sample}$). Once these key parameters are determined, the molar loading (i.e. fraction of a species' mass fraction and its molar weight) is used to calculate the key performance indicators such as the calcination degree (eq. 4.9) and the recarbonation degree (eq. 4.10). The term recarbonation degree is used instead of the CO_2 carrying capacity to assess the CO_2 uptake potential based on the available CaO since the entrained flow calcined samples were only partially calcined. The denominator in equation 4.10 represents the available molar amount of CaO, whereas the numerator corresponds to the molar amount of $CaCO_3$ formed or CO_2 absorbed during carbonation.

$$x_{CO_2,EF} = \frac{\Delta M_{calc}}{M_{0,sample}} \tag{4.6}$$

$$x_{CaO,EF} = x_{CaO,raw\,meal} \cdot \frac{1 - x_{CO_2,EF}}{1 - x_{CO_2,raw\,meal}} \tag{4.7}$$

$$x_{CO_2,carb} = \frac{\Delta M_{EF}}{M_{0,sample}} = \frac{M_{0,sample} - M_{Ref,carb}}{M_{0,Sample}} \tag{4.8}$$

$$X_{calc} = 1 - \frac{\dfrac{x_{CO_2,carb}}{\tilde{M}_{CO_2}}}{\dfrac{x_{CaO,EF}}{\tilde{M}_{CaO}}} \tag{4.9}$$

$$X_{recarb} = \frac{\dfrac{x_{CO_2,carb}}{\tilde{M}_{CO_2}}}{\dfrac{x_{CaO,EF}}{\tilde{M}_{CaO}} - \left(\dfrac{x_{CO_2,EF}}{\tilde{M}_{CO_2}}\right)} \tag{4.10}$$

Analysing the influence of CO_2 partial pressure and carbonation temperature on the sorbent conversion, the sorbent sample starts calcining during heat up to the respective carbonation temperature. In these cases, the recarbonation degree was calculated by equation 4.11 assuming that the CaO formed during heat up has the same recarbonation degree as the CaO formed during entrained flow calcination. The carbonation reaction rate in the chemically controlled regime is again determined by linear regression around the inflection point of the TGA signal to minimise potential mass transfer limitations.

$$X_{recarb} = \frac{\dfrac{x_{CO_2,carb}}{\tilde{M}_{CO_2}}}{\dfrac{x_{CaO,EF}}{\tilde{M}_{CaO}} - \left(\dfrac{x_{CO_2,EF}}{\tilde{M}_{CO_2}} - \dfrac{x_{CO_2,heat\text{-}up}}{\tilde{M}_{CO_2}}\right)} \tag{4.11}$$

Chapter 5

Results - sorbent screening

In this chapter the results concerning the CO_2 uptake potential of different sorbents are presented. Various limestones and raw meals as well as raw meal components were analysed using thermogravimetric analysis (TGA). Initially, the carbonation conversion of the various limestones are briefly assessed. Subsequently, the CO_2 uptake potential of different cement raw meals including their raw components is presented.

5.1 Limestone

Five different limestones originating from various geological sites have been assessed by the reference analysis cycle described in section 4.1. Additionally, three particle size cuts of the Vernasca limestone were investigated, as well as two size cuts of the Lhoist limestone from Messinghausen (RK). The Vernasca limestone is a limestone rich raw meal component that accounts for approx. $0.28\,\mathrm{kg\,kg^{-1}}$ of the Vernasca raw meal and contributes to 34 % of the raw meal's overall calcium content. The CO_2 carrying capacities of the different limestones are depicted in figure 5.1, while the cyclic sorbent activity of three particle size cuts of the Vernasca limestone are presented in figure 5.2. The characteristic decay of the limestone's CO_2 carrying capacity is evident for all investigated limestones regardless of their origin or particle size. While in the initial carbonation cycle around two thirds of the limestone's calcium oxide is carbonated, the conversion decreases exponentially towards a threshold (i.e. residual CO_2 carrying capacity). After 20 cycles a CO_2 carrying capacity of approx. $0.12\,\mathrm{mol\,mol^{-1}}$ is obtained. No significant influence of the sorbent's particle size on its CO_2 carrying capacity can be observed in the investigated size range. The obtained CO_2 carrying capacity of the individual limestones can be adequately described by the model developed by Grasa et al. [76] with a deactivation constant (k) in the range of 0.47 to 0.80

© The Author(s), under exclusive license to
Springer Fachmedien Wiesbaden GmbH, part of Springer Nature 2022
M. Hornberger, *Experimental Investigation of Calcium Looping CO2
Capture for Application in Cement Plants*,
https://doi.org/10.1007/978-3-658-39248-2_5

and a residual activity between $0.04\,\mathrm{mol\,mol^{-1}}$ and $0.08\,\mathrm{mol\,mol^{-1}}$. Fitting the obtained CO_2 carrying capacity to all sorbents yield a deactivation constant of 0.60 ± 0.03 and a residual activity of $0.058\,\mathrm{mol\,mol^{-1}} \pm 0.007\,\mathrm{mol\,mol^{-1}}$. The characteristic deactivation parameters of the investigated limestones obtained by fitting the experiments to the model of Grasa are summarised in table 5.1. Such a deactivation behaviour has widely been reported for pure limestones commonly assessed for calcium looping purposes [74]. Deactivation constants reported in the literature range from 0.33 to 1.96 depending on the calcination conditions, whereas the residual activities reported range from $0.04\,\mathrm{mol\,mol^{-1}}$ to $0.18\,\mathrm{mol\,mol^{-1}}$. Grasa et al. [76] concluded that the CO_2 carrying capacity of most limestones can be described using an universal deactivation constant of 0.52 with a residual activity of $0.075\,\mathrm{mol\,mol^{-1}}$ which is also in very good agreement with the CO_2 carrying capacity obtained for the limestones in this work. The consistent deactivation characteristics of the different limestone qualities can most likely be attributed to their high purity. The high purity or rather the absence of impurities reduces the possibility of side reactions. Furthermore, an enhancement of the sorbent's sintering characteristics is avoided as the calcium mobility is not promoted by impurities [34]. Consequently, the sorbents yield similar deactivation characteristics mostly associated with sintering.

Based on the limestone sorbent characterisation it can be anticipated that a tail-end fluidised bed calcium looping system operated with a relatively pure limestone fraction of the raw meal will perform similarly to a power plant application, but with the benefit of a more active sorbent since the possibility of sorbent reutilisation will allow the operation with higher sorbent make-up flows.

Table 5.1: Summary of characteristic deactivation fitted to the model of Grasa et al. [76] for the five instigated limestones.

Limestone	Deactivation constant k	Residual sorbent activity X_r in $\mathrm{mol\,mol^{-1}}$
EnBW	0.4682	0.042
RK	0.5891	0.050
Vernasca	0.6282	0.076
Verdal	0.6018	0.057
Saabar	0.7977	0.056
All	0.6061	0.0569

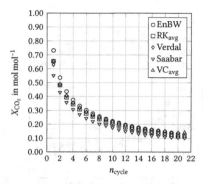

Figure 5.1: CO_2 carrying capacity for various limestones. Calcination at 850 °C in N_2 for 10 min, carbonation at 650 °C with $0.15\,m^3\,m^{-3}$ CO_2 in N_2 for 15 min.

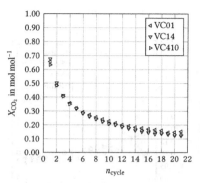

Figure 5.2: CO_2 carrying capacity for different particle fractions of Vernasca limestone (VC). Calcination at 850 °C in N_2 for 10 min, carbonation at 650 °C with $0.15\,m^3\,m^{-3}$ CO_2 in N_2 for 15 min.

5.2 Raw meal

In this section, a comprehensive study conducted with five different raw meals, including their raw components, is presented. The analysis focuses on the impacts on the CO_2 carrying capacity using the TGA routines described in section 4.1. The CO_2 carrying capacity of the different raw meal mixtures are presented in figure 5.3 categorised by the respective TGA routine (i.e. calcination/carbonation conditions). In figure 5.4 their CO_2 carrying capacities are sorted by the individual raw meals to facilitate the assessment of a particular change of the reaction conditions on the raw meal's CO_2 carrying capacity. All raw meals show an exponential decay towards their residual threshold activity similar to limestones but with a stronger initial reduction of the CO_2 carrying capacity. For the reference case (i.e. dry air calcination), the CO_2 carrying capacity after the first calcination drops to values between $0.17\,mol\,mol^{-1}$ and $0.22\,mol\,mol^{-1}$ for all raw meals with an exception of the Vernasca raw meal. The Vernasca raw meal exhibits an initial CO_2 carrying capacity of $0.45\,mol\,mol^{-1}$ (figure 5.3a). Eventually, the sorbent activity converges towards the sorbent activity of the other raw meals (after the sixth cycle). Despite the deviation in the initial CO_2 carrying capacity all investigated raw meals decay towards a residual CO_2 carrying capacity of around $0.1\,mol\,mol^{-1}$, which is also characteristic for limestone based sorbents

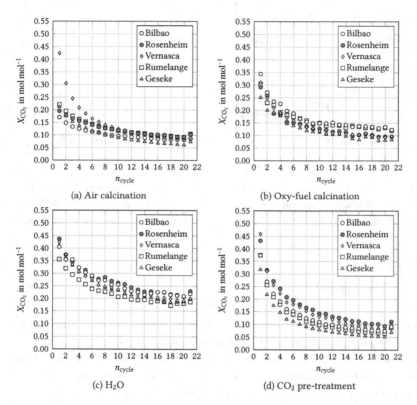

(a) Air calcination

(b) Oxy-fuel calcination

(c) H₂O

(d) CO₂ pre-treatment

Figure 5.3: CO₂ carrying capacity (X_{CO_2}) vs. number of cycles (n_{cycle}) of various raw meals analysed for four different TGA routines.

(a) Air calcination: T_{calc} = 850 °C, t_{calc} = 10 min, $y_{N_2,calc}$ = 1.00 m³ m⁻³, T_{carb} = 650 °C, t_{carb} = 10 min, $y_{CO_2,carb}$ = 0.15 m³ m⁻³;

(b) oxy-fuel calcination: T_{calc} = 920 °C, t_{calc} = 10 min, $y_{CO_2,calc}$ = 0.85 m³ m⁻³, T_{carb} = 650 °C, t_{carb} = 10 min, $y_{CO_2,carb}$ = 0.15 m³ m⁻³, y_{H_2O} = 0.15 m³ m⁻³;

(c) H₂O: air calcination programme with 0.15 m³ m⁻³ of steam;

(d) CO₂ pre-treatment: initial heat up to 850 °C and 'tempering' for 20 min in pure CO₂ followed by the air calcination programme.

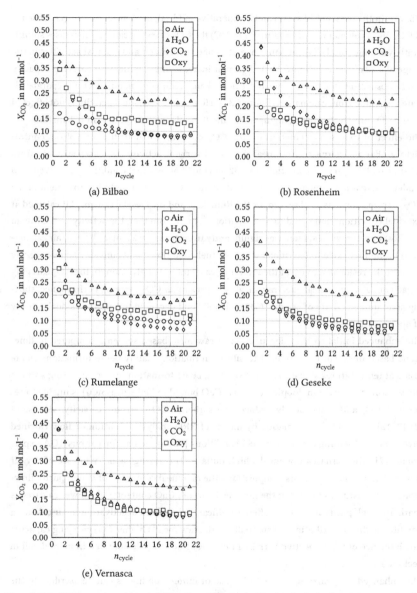

Figure 5.4: CO_2 carrying capacity (X_{CO_2}) vs. number of cycles (n_{cycle}) of various raw meals analysed for four different TGA routines.

[76, 77]. In the presence of steam and comparative mild calcination conditions (i.e. absence of CO_2 and a calcination temperature of 860 °C) the CO_2 carrying capacity increases significantly yielding initial CO_2 carrying capacities of $0.5 \, mol \, mol^{-1}$ and residual activities in the range of $0.22 \, mol \, mol^{-1}$ (figure 5.3c). A similar increase of the initial CO_2 carrying capacity can be observed when the raw meals are heated and 'tempered' in pure CO_2 to separate silicate formation from calcination (figure 5.3d). However, the residual activity is not affected by the CO_2 pre-treatment averaging at around $0.1 \, mol \, mol^{-1}$ as for dry air calcination.

The effect of oxy-fuel calcination on the CO_2 carrying capacity (figure 5.3b) is contradictory. For the Vernasca raw meal the initial sorbent activity decreases strongly from $0.45 \, mol \, mol^{-1}$ at dry air conditions to $0.32 \, mol \, mol^{-1}$ at oxy-fuel conditions, followed by a milder deactivation compared to dry air calcination. Contrary, a slight improvement of the CO_2 carrying capacity is observed for the Rumelage and the Geseke raw meal if calcined at oxy-fuel conditions compared to air conditions. Apart from that, the activity of the Bilbao and Rosenheim raw meal increases significantly from around $0.2 \, mol \, mol^{-1}$ (dry air calcination) in the first cycle to $0.35 \, mol \, mol^{-1}$ (oxy-fuel conditions) and converges subsequently towards the residual activity of the reference case. Interestingly enough, the CO_2 carrying capacity of all oxy-fuel calcined raw meals show a similar CO_2 carrying performance, starting at an initial activity of approx. $0.35 \, mol \, mol^{-1}$ and decaying towards a residual activity of approx. $0.1 \, mol \, mol^{-1}$.

The enhanced initial sorbent deactivation of raw meal based sorbents compared to limestone based ones can be attributed to silicate formation, mainly belite formation. Belite arises at temperatures above 700 °C [113]. Thus, belite formation can occur during sorbent calcination in the calcium looping calciner. CaO bound in belite (Ca_2SiO_4) is unable to react with CO_2 and consequently, reduces the sorbent's CO_2 carrying capacity. Alonso et al. [7, 12] verified belite formation by means of XRD analysis of entrained flow calcined raw meal representing air and oxy-fuel conditions [12] as well as for various TGA experiments [7]. The Vernasca raw meal's high initial CO_2 carrying capacity in case of dry air calcination originates from its composition. The raw meal comprises of a high purity limestone that constitutes to 34 % of the raw meals overall CaO content and a lower calcareous marl. The high purity limestone suffers significantly less from deactivation in air calcining conditions, thus contributing substantially to the raw meal's higher CO_2 carrying capacity. The influence of the respective Vernasca raw meal fractions is elaborated in more detail in section 5.3.

The enhanced CO_2 carrying capacity in case of humid conditions can be ascribed to the promoting effect of the presence of steam. It is widely agreed that steam has an enhancing effect on the carbonation reaction [56–58, 120]. The migration of ions such as O^{2-} and Ca^{2+} is facilitated due to the presence of steam promoting the coalescence and growth of product layer islands that preserve more of the reactive surface [120]. The improved CO_2 carrying capacity initially obtained for sorbent 'tempered' in CO_2 at 850 °C is somewhat against expectations. Generally, prolonged residence times at high temperatures and/or high CO_2 concentrations lead to severe sintering [32, 34, 67]. However, calcination could not occur due to equilibrium constraints as a consequence, only silicate formation could arise. Silicate formation has been confirmed by means of mass loss in the holding step. It is hypothesised that a more stable pore network is formed that enables enhanced access to the sorbent's internal surface area. The subsequent decay in sorbent activity can be attributed to the gradual loss of accessible surface area. A similar phenomenon was observed for some limestone based sorbents. Initial high temperature sintering produced a sort of scaffold that prevented the particle pore system from collapsing and avoided particle densification [122]. This endorses the hypothesis, that a supportive structure is formed by belite formation allowing enhanced access to the inner surface of the sorbent particle. The hypothesis is further supported by the fact that the Rosenheim raw meal yielded a significant increase in BET surface area when calcined in CO_2 rich atmosphere compared to N_2. However, the Bilbao raw meal showed neither a significant improvement nor a decline of the BET surface when calcined in CO_2. Nonetheless, it has to be highlighted that the sorbent morphology might be altered if belite formation occurs alongside calcination.

The CO_2 carrying capacity of all tested raw meals can be adequately described at oxy-fuel conditions, using the model of Grasa et al. [76] developed for limestones, with a deactivation constant of 3.29 and a residual activity of 0.10 mol mol^{-1}. Ortiz et al. [133] proposed an equation following the work of Grasa et al. [76] and Valverde [171] to account for a strong initial deactivation associated with enhanced sintering (eq. 5.1). Fitting the CO_2 carrying capacity obtained for oxy-fuel conditions to the model of Ortiz to account for the strong initial deactivation of the belite formation yields an initial CO_2 carrying capacity (X_1) of 0.30 mol mol^{-1}, a residual CO_2 carrying capacity (X_r) of 0.076 mol mol^{-1} and a deactivation constant of 0.44. The high difference between the obtained deactivation constants can be ascribed to the fact that the deactivation constant of the Ortiz model describes the deactivation subsequent to the initial deactivation, whereas the deactivation constant of the model of Grasa needs to predict the strong initial decay in the first cycle as well as the subsequent decrease. The two models describing the raw meal's CO_2 carrying capacity in case of oxy-fuel calcination are depicted in figure 5.5. The characteristic deactivation values of all raw meals and TGA analyses are summarised in table 5.2.

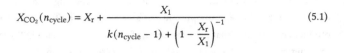

$$X_{CO_2}(n_{cycle}) = X_r + \cfrac{X_1}{k(n_{cycle} - 1) + \left(1 - \cfrac{X_r}{X_1}\right)^{-1}} \tag{5.1}$$

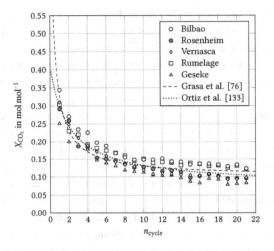

Figure 5.5: CO_2 carrying capacity (X_{CO_2}) vs. number of calcination and carbonation cycles (n_{cycle}) for oxy-fuel calcined raw meals including the model lines of Grasa et al. [76] and Ortiz et al. [133].

Additional experiments with a reduced calcination time of 1 min (instead of 10 min) have been conducted in order to limit the formation of belite and assess operation conditions more representative for entrained flow calcination in an integrated entrained flow calcium looping system. The experiments with a reduced calcination time are presented in figure 5.6a for dry air calcination conditions and in figure 5.6b for oxy-fuel conditions. In case of oxy-fuel conditions, the calcination time was insufficient to fully calcine the sorbent in the first cycle if the experiments were conducted with a sample mass of 12 mg. The 12 mg experiments (symbols filled in white) were used to assess the raw meal's CO_2 carrying capacity for higher calcination and carbonation cycles (i.e. low CO_2 carrying capacities) maintaining an adequate signal to noise ratio, whereas additional runs with a reduced sample mass of 5 mg experiments (symbols filled in grey) were used to assess the raw meal's CO_2 carrying capacity in the first three cycles. Within the second cycle the 5 mg experiments and the 12 mg experiments converge towards the same sorbent activity indicating the validity of

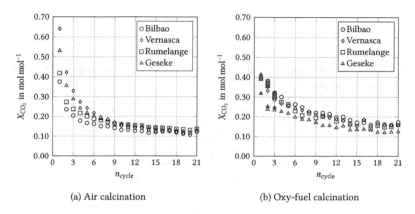

(a) Air calcination   (b) Oxy-fuel calcination

Figure 5.6: CO_2 carrying capacity (X_{CO_2}) vs. number of calcination and carbonation cycles (n) for calcination times of 1 min at air conditions (a) and oxy-fuel conditions (b).

the approach. At air calcining conditions the experiments with a reduced calcination time yield significant higher sorbent activities for all investigated raw meals compared to the 10 min calcination experiments. For instance, the Bilbao's CO_2 carrying capacity increased from $0.18\,mol\,mol^{-1}$ to $0.42\,mol\,mol^{-1}$ in the first cycle, whereas the Vernasca raw meal's CO_2 carrying capacity increased from $0.42\,mol\,mol^{-1}$ to $0.64\,mol\,mol^{-1}$. The initial CO_2 carrying capacity of the Rumelange and Geseke raw meal increased from $0.23\,mol\,mol^{-1}$ to $0.44\,mol\,mol^{-1}$ and $0.57\,mol\,mol^{-1}$, respectively. For oxy-fuel experiments a slight improvement of the CO_2 carrying capacity from around $0.30\,mol\,mol^{-1}$ to $0.38\,mol\,mol^{-1}$ is observed when the calcination time is reduced from 10 min to 1 min. Despite the initial improvement, the CO_2 carrying capacity converged towards $0.1\,mol\,mol^{-1}$ around the 20th cycle for all raw meals regardless of the reaction atmosphere or calcination time. The CO_2 carrying capacities of the oxy-fuel experiments with a calcination time of 1 min can also be adequately described using a single model line. Fitting all experiments to the model of Grasa et al. [76] yield a deactivation constant of 2.27 with a residual sorbent activity of $0.1452\,mol\,mol^{-1}$. Using the model of Ortiz et al. [133], the CO_2 carrying capacity can be described with an initial CO_2 carrying capacity of $0.3855\,mol\,mol^{-1}$, a deactivation constant of 0.28 and a residual CO_2 carrying capacity of $0.0910\,mol\,mol^{-1}$. Each raw meal's individual model parameters can be obtained from table 5.2.

Table 5.2: Summary of characteristic deactivation values for the model proposed by Grasa et al. [76] as well as Ortiz et al. [133].

		Grasa et al. [76]		Ortiz et al. [133].		
Raw meal	TGA run	k	X_r	k	X_r	X_1
Bilbao	Air	8.56	0.0841	0.37	0.0625	0.1697
	Oxy	2.88	0.1190	0.54	0.0992	0.3431
	CO_2	1.88	0.0572	0.47	0.0398	0.3724
	H_2O	2.87	0.2115	0.26	0.1478	0.4042
	Air 1 min	2.69	0.0992	1.22	0.1047	0.3724
	Oxy 1 min	2.03	0.1584	0.23	0.0893	0.4100
Rosenheim	Air	7.10	0.1015	0.19	0.0546	0.1960
	Oxy	3.18	0.0915	0.44	0.0668	0.2968
	CO_2	1.53	0.0797	0.39	0.0515	0.4255
	H_2O	2.55	0.2135	0.35	0.1656	0.4301
Vernasca	Air	1.54	0.0604	0.49	0.0460	0.4227
	Oxy	2.81	0.0869	0.40	0.0591	0.3108
	CO_2	1.36	0.0583	0.50	0.0467	0.4526
	H_2O	2.30	0.1889	0.36	0.1440	0.4265
	Air 1 min	0.77	0.0461	0.66	0.0672	0.6389
	Oxy 1 min	2.09	0.1481	0.35	0.1057	0.4092
Rumelage	Air	5.42	0.0978	0.26	0.0594	0.2217
	Oxy	3.90	0.1220	0.62	0.1054	0.2999
	CO_2	1.87	0.0475	0.54	0.0362	0.3728
	H_2O	3.35	0.1783	0.33	0.1323	0.3577
	Air 1 min	2.29	0.1168	1.03	0.1195	0.4138
	Oxy 1 min	2.08	0.1452	0.30	0.0938	0.4017
Geseke	Air	4.58	0.0653	0.26	0.0285	0.2108
	Oxy	3.91	0.0838	0.27	0.0434	0.2484
	CO_2	2.34	0.0377	0.54	0.0263	0.3169
	H_2O	2.34	0.1827	0.28	0.1244	0.4116
	Air 1 min	1.25	0.0877	0.78	0.0963	0.5296
	Oxy 1 min	3.30	0.1267	0.22	0.0680	0.3055
All	Oxy	3.29	0.1008	0.44	0.07551	0.2996
	Oxy 1 min	2.27	0.1452	0.28	0.09096	0.3855

Due to the slow nature of solid-solid reactions, the belite formation is expected to decrease with decreasing residence time at calcining conditions resulting in a more active raw meal based sorbent. The results further indicate that silicate formation is more pronounced at operation conditions favouring calcination as the reduction of calcination time yields a significant improvement of the sorbent's CO_2 carrying capacity at air calcination conditions. Assumingly, the improved calcination kinetics lead to an increasing availability of CaO to form belite via the indirect pathway (eq. 2.17). Therefore, belite formation progresses severely with increasing residence time. The minor improvement of the CO_2 carrying capacity when reducing the calcination time in case of oxy-fuel calcination indicates that the silicate formation is inhibited by the higher CO_2 partial pressure. The high CO_2 partial pressure directly inhibits silicate formation via the direct pathway and in an indirect way by impeding the calcination reaction (indirect pathway). Eventually, belite formation will gradually complete with increasing cycles yielding the same residual activity regardless of the calcination time.

5.3 Raw meal components

Cement raw meals are often blended by two or more raw meal components such as marlstones, limestones or corrective additives of various origins. The respective raw meal components of the Geseke and Rumelage raw meals have been investigated regarding their CO_2 carrying performance at dry air and oxy-fuel conditions. Furthermore, the raw meal components of the Vernasca raw meal have been investigated using all TGA routines previously described in section 4.1. The chemical compositions of the raw meal components can be derived from table 4.2 in section 4.1. The Bilbao and Rosenheim raw meals represent marl type raw meals consisting of a single marlstone with a suitable composition to produce clinker. Consequently, no raw meal components can be tested for these raw meals. It is evident from figure 5.8 (Geseke) and figure 5.7 (Rumelange) as well as figure 5.9 (Vernasca) that the various raw meal components show a certain diversity in their CO_2 capture performance depending on their chemical composition. Generally, it appears that the main component affecting the raw meal component's CO_2 carrying capacity is silicon as it allows the formation of belite which binds CaO making it unable to react with CO_2. This fact can clearly be seen by comparing the Rumelange's limestone fraction (R_{Lime}, $\gamma_{SiO_2} = 0.08$ kg kg^{-1}) with its higher siliceous components ($\gamma_{SiO_2} = 0.16 \ldots 0.19$ kg kg^{-1}) and also when comparing the North and South fraction of the Geseke raw meal with its corrective marl. Nonetheless, a precise prediction of the sorbent's CO_2 carrying capacity based solely on its composition is

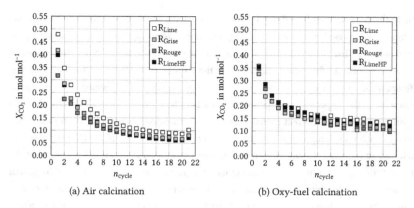

(a) Air calcination   (b) Oxy-fuel calcination

Figure 5.7: CO_2 carrying capacity (X_{CO_2}) vs. number of calcination and carbonation cycles (n_{cycle}) for the Rumelange (R) raw meal components.

insufficient as the sorbent activity is also influenced by other characteristics such as sorbent morphology or calcium to silicon distribution. This circumstance can be seen from the two marly limestone fractions of the Geseke raw meal. Both the Geseke North and South component have a SiO_2 content of approx. $0.14\,kg\,kg^{-1}$. However, the South component yields a higher initial sorbent activity of approx. $0.34\,mol\,mol^{-1}$ at air conditions and $0.28\,mol\,mol^{-1}$ for oxy-fuel conditions compared to $0.27\,mol\,mol^{-1}$ and $0.23\,mol\,mol^{-1}$ of the North fraction. Assuming that belite formation is majorly responsible for the sorbent deactivation and correcting the amount of CaO available for CO_2 capture accordingly, clearly indicate that not all SiO_2 is reacting with the CaO. It can be anticipated that belite formation occurs at grain boundaries of siliceous and calcareous grains but does not migrate into the grain itself given the slow nature of solid-solid reactions. Thus, the Ca-Si distribution is a crucial parameter for the deactivation of raw meal based calcium looping sorbents. Alonso et al. [7] also highlighted the importance of Ca-Si distribution on sorbent deactivation and concluded that homogeneous aggregation of calcium and silicon leads to a more severe initial sorbent deactivation. Furthermore, the probability of the proximity of calcium and silicon increases with increasing SiO_2 content.

Potential interactions between the different raw meal components have been assessed using the Vernasca limestone fraction since a precise composition of the raw meal mixture was known. The raw meal composes to $0.72\,kg\,kg^{-1}$ of a lower calcareous marl and to $0.28\,kg\,kg^{-1}$ of a high grade limestone with minor SiO_2 impurities in the range of $0.03\,kg\,kg^{-1}$.

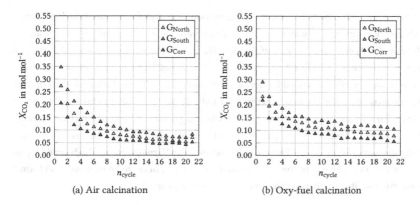

(a) Air calcination   (b) Oxy-fuel calcination

Figure 5.8: CO_2 carrying capacity (X_{CO_2}) vs. number of calcination and carbonation cycles (n_{cycle}) for the Geseke (G) raw meal components.

The results are presented in figure 5.9 again as CO_2 carrying capacity against the cycle number. Additionally, the CO_2 carrying capacities of the Vernasca raw meal calculated by means of superposition of the various raw component's molar CaO content are presented. Very good agreement between the CO_2 carrying capacities obtained by TGA analysis and those calculated using the superposition principle can be observed for most analyses. Minor over-prediction of the CO_2 carrying capacity of $0.1 \, mol \, mol^{-1}$, that diminishes with subsequent cycles, occurs in case of pre-silicate formation (i.e. heat up and 'tempering' in pure CO_2, figure 5.9d). The overprediction might be associated with the prolonged residence time at 850 °C of 20 min giving the inter-particle reactions more time to progress. Based on these results it can be anticipated that no substantial interaction between the raw meal components occurs at calcium looping operation conditions. For one thing, the temperature in the oxy-fuel calciner is insufficient to enable partial melting of the particles allowing the raw meal components to rapidly react. Furthermore, the residence time in both entrained flow and fluidised bed calciners is too short for a substantial progress of inter-particle reactions. Overall, the limestone fraction showed an increased CO_2 carrying capacity in the presence of steam (figure 5.9c). Compared to air calcination, the CO_2 carrying capacity at oxy-fuel conditions was hardly affected indicating that deterioration by sintering and the improvement due to the presence of steam compensate each other (figure 5.9b). The pre-treatment in pure CO_2 did not affected the limestone's activity due to the fact that neither belite formation (due to the absence of SiO_2) nor sintering (due to non-calcining conditions) could occur (figure 5.9d). The reduction of the calcination time slightly increased the limestone's CO_2 carrying capacity indicating a reduced deactivation due to sintering (figure 5.9e and f). The initial CO_2 carrying capacity of the marl fraction of the Vernasca raw meal decreased from $0.30\,mol\,mol^{-1}$ at dry air conditions to $0.17\,mol\,mol^{-1}$ at oxy-fuel conditions. Also, the residual activity decreased by approx. $0.02\,mol\,mol^{-1}$. The reduced calcination time significantly increases the marl's activity for both air and oxy-fuel conditions (figure 5.9e and f). The CO_2 pre-treatment of the marl fraction significantly increased its CO_2 carrying capacity (figure 5.9d) while the presence of steam minorly improved the marl's CO_2 carrying capacity (figure 5.9c). These results indicate the Vernasca marl's high tendency to from belite. For such marls, the CO_2 pre-treatment step appears to be reasonable to form a supporting scaffold increasing the marl's CO_2 carrying capacity.

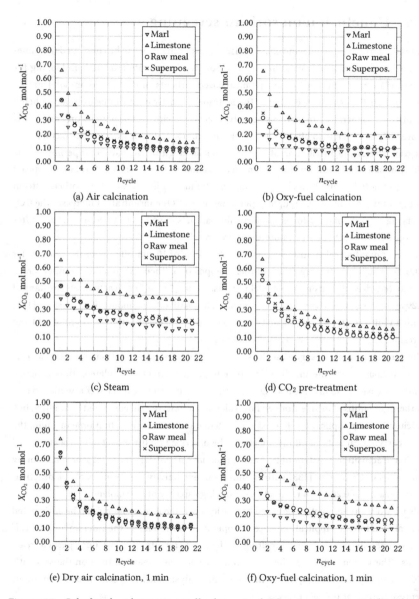

Figure 5.9: Calculated and experimentally determined CO_2 carrying capacity (X_{CO_2}) vs. number of calcination and carbonation cycles (n_{cycle}) of the Vernasca raw meal and its raw meal components.

5.4   Conclusion on sorbent screening

From the sorbent screening and characterisation experiments it can be concluded that limestone based sorbents behave as anticipated from the investigation of calcium looping for
power plant application, whereas raw meal based sorbents suffer a severe initial degradation in their CO_2 carrying capacity. The additional deactivation can be ascribed to belite
formation. Generally, the sorbent activity decreased with increasing silicon content, as
more SiO_2 is available to bind CaO and form calcium silicate. However, other factors such
as grain size or the distribution of calcium and silicon within the particle superimpose the
effect of silicon content on the sorbent deactivation. Calcium silicate formation appears
to be promoted by more calcining conditions. Similar to limestone based sorbents steam
improves the raw meal's CO_2 uptake potential. High CO_2 concentration increases the CO_2
carrying capacity of marlstones as silicate formation is impeded, whereas limestones suffer
a more severe deactivation as sintering is enhanced.

In case of oxy-fuel calcination with a calcination time of 10 min, all raw meals yield similar
CO_2 carrying capacities regardless of their composition. Their CO_2 uptake potential can be
adequately described using the model of Grasa et al. [76] with a deactivation constant of
3.29 and a residual activity of $0.10\,mol\,mol^{-1}$. Alternatively, the model of Ortiz et al. [133]
with an initial CO_2 carrying capacity of $0.30\,mol\,mol^{-1}$, a deactivation constant of 0.44 and
a residual activity of $0.076\,mol\,mol^{-1}$ can be applied.

Shorter calcination times increase the sorbent activity since deactivation processes, such as
belite formation or sintering, have less time to progress and be completed. Thus, a significant increase in the raw meal's CO_2 carrying capacity can be expected if raw meal based
sorbents are utilised in an entrained flow calcium looping system compared to fluidised bed
calcium looping application. For oxy-fuel calcination with a calcination time of 1 min, the
CO_2 carrying capacity of all raw meals can be described by the model of Grasa et al. [76]
with a deactivation constant of 2.27 and a residual sorbent activity of $0.15\,mol\,mol^{-1}$ or by
the model of Ortiz et al. [133] with an initial sorbent activity of $0.39\,mol\,mol^{-1}$, a deactivation constant of 0.28 and a residual CO_2 carrying capacity of $0.09\,mol\,mol^{-1}$.

Also, hardly any interaction between the various raw meal components can be anticipated
for calcium looping application since the operation conditions in the calcium looping calciner generally prevent partial melting of the solids required for fast inter-particle reactions. The various effects of the reaction atmosphere and reaction time on the sorbent's
CO_2 carrying capacity emphasise that sorbent analysis shall be performed at conditions
representative for the real process.

Chapter 6

Results - fluidised bed calcium looping

In this section the results concerning calcium looping CO_2 capture from cement plants using fluidised bed reactors are presented. The results have been obtained using University of Stuttgart's fluidised bed pilot plant and have been partially published in two journal articles [85, 86]. The initial section addresses process operation and evaluation of the experiments while the later part covers the process' CO_2 capture performance and the performance of the oxy-fuel calciner in more detail. The influence of high integration levels on the performance of the calcination and carbonation reactor was assessed at different calcination and carbonation temperatures with a flue gas CO_2 concentration of $0.15\,m^3\,m^{-3}$ using the CFB-CFB configuration. Since the maximum feeding rate of limestone was limited to $50\,kg\,h^{-1}$ by the feeder's rotational speed which limited the make-up rate to approx. $0.6\,mol\,mol^{-1}$, the BFB configuration was employed to assess even higher make-up ratios of $0.9\,mol\,mol^{-1}$. Lower integration levels have been assessed using the CFB-CFB configuration by testing a variety of make-up rates ($15\,kg\,h^{-1}$ to $45\,kg\,h^{-1}$) and CO_2 concentrations ($0.20\,m^3\,m^{-3}$ to $0.35\,m^3\,m^{-3}$) at a carbonation temperature of approx. 650 °C, known to be beneficial for less active sorbent, and a calcination temperature of approx. 920 °C.

© The Author(s), under exclusive license to
Springer Fachmedien Wiesbaden GmbH, part of Springer Nature 2022
M. Hornberger, *Experimental Investigation of Calcium Looping CO2 Capture for Application in Cement Plants*,
https://doi.org/10.1007/978-3-658-39248-2_6

6.1   Carbon mass balance of conducted experiments

The carbon balance of the conducted experiments has been performed using the most reliable measurements available such as gas analysis, volume flows from mass flow controllers (MFC) and the mass flows of the gravimetrically controlled dosing system. Figure 6.1 depicts various balance boundaries used for the assessment of the carbon balance.
Initially, the carbonator's CO_2 capture efficiency ($E_{CO_2,carb}$) or absorbed amount of CO_2

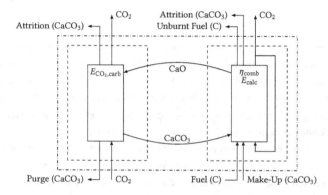

Figure 6.1: Schematic of the system boundaries used to perform the carbon balance.

($E_{CO_2,carb} \cdot \dot{V}_{CO_2,carb,in}$) was determined by solving equation 6.1 using the CO_2 inlet volume
flow measured by a MFC ($\dot{V}_{CO_2,carb,in}$) as well as the online gas measurements at the carbonator inlet ($y_{CO_2,carb,in}$) and outlet ($y_{CO_2,carb,out}$). This method was chosen since the volume
flow measured by impeller anemometer is less reliable due to the fact that particle loaded
gas is measured and the flow profile of the flue gas is strongly influenced by the reduction
of the cross section of the impeller anemometer in the comparatively small flue gas duct.
Subsequently, the amount of CO_2 transferred to the calciner in form of $CaCO_3$ ($\dot{N}_{CO_2,loop}$)
was determined by subtracting the amount of CO_2 purged from the carbonator by means of
$CaCO_3$ ($\dot{N}_{CO_2,carb,purge}$) as well as the loss of CO_2 by sorbent attrition ($\dot{N}_{CO_2,carb,attr}$). This approach eradicates uncertainties imposed on the system by the determination of circulating
mass flow as well as the accuracy of the impeller anemometer, but simultaneously resulting in the full closure of the mass balance. The assumption that CO_2 removed from the gas
phase in the carbonator is transferred to the calciner can be justified since accumulation of
solids can hardly occur in the transfer line of the pilot facility.

$$y_{CO_2,carb,out} = \frac{\dot{V}_{CO_2,carb,in} \cdot (1 - E_{CO_2,carb})}{\frac{\dot{V}_{CO_2,carb,in}}{y_{CO_2,carb,in}} - E_{CO_2,carb} \cdot \dot{V}_{CO_2,carb,in}} \tag{6.1}$$

The CO_2 in the calciner originates from (i) the combustion of coal, (ii) the calcination of (a) circulating sorbent and (b) sorbent make-up as well as (iii) the recirculated flue gas to control the temperature. In a first step the calcination efficiency (E_{calc}) was determined by equation 2.25 using solid sample analysis. Subsequently, fuel conversion (η_{comb}) was determined by solving equation 6.2, performing a combustion calculation with superimposed calcination to determine the total flue gas flow. Since the oxygen concentration was measured on a dry basis by NDIR gas analysis this calculation was also performed on a dry basis. Assessing the calciner's mass balance using efficiencies also results in full closure of its mass balance since unconverted material is leaving the reactor by definition. Consequently, the determined CO_2 concentration, steam concentration (if available) and flue gas volume flow have been used to evaluate the accuracy of the conducted experiments.

$$y_{O_2,calc,out} = \frac{\dot{V}_{O_2,MFC} - \eta_{comb} \cdot \dot{M}_{coal} \cdot (1 - \gamma_{H_2O}) \cdot \mu_{O_2} \cdot \frac{\tilde{V}}{\tilde{M}_{O_2}}}{\dot{V}_{tot,calc,out}(\eta_{comb}, E_{calc})} \tag{6.2}$$

Figure 6.2 presents the agreement between the measured and calculated CO_2 and steam concentrations, as well as the flue gas volume flows. Deviation between the calculated and measured CO_2 concentration averaged at around $0.015\,m^3\,m^{-3}$ with an average CO_2 concentration of approx. $0.81\,m^3\,m^{-3}$ (on a dry basis), whereas the deviation of the determined and measured volume flow averaged around $20\,m^3\,h^{-1}$ with an average volume flow in the range of $120\,m^3\,h^{-1}$. Steam measurement by jet impact psychrometry is susceptible to malfunction. Hence, reliable steam measurements were not available for all conducted experiments. Deviation of the determined and measured steam concentration averaged to $0.0095\,m^3\,m^{-3}$ with an overall steam concentration of $0.20\,m^3\,m^{-3}$ in the initial fifteen experiments and $0.17\,m^3\,m^{-3}$ for experiment 45 to 57. The comparatively high deviation of the measured and determined volume flow with the tendency to overpredict the measured volume flow can be attributed to the utilised measurement device (i.e. impeller anemometer). Measuring the velocity of a flow medium, impeller aneometers are susceptible to error in the flow profile. The installation of the impeller aneometer reduces the pipe's cross-section and consequently affects the flue gas' flow profile. This effect can be neglected for large pipe diameters but is more pronounced for small pipe diameters like the calciner's flue gas duct. The stronger deviation of the measured and solved CO_2 concentrations for the experiments 16 to 22 can be ascribed to false air ingress in the calciner's flue gas duct. The circumstance that an increased calciner inventory was required to enable operation of the BFB carbonator with a reference solid inventory of 80 kg in slight overpressure, led to negative pressure in the calciner's flue gas duct causing the false air ingress via the bin of the protective cyclone.

Overall, sufficient agreement between the measured and calculated process variables has been achieved using standard combustion calculation with superimposed calcination. The agreement emphasises the reliability of the conducted experiments.

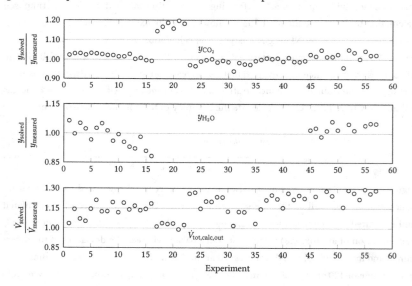

Figure 6.2: Comparison of calculated ($_{solved}$) and measured ($_{measured}$) CO_2 concentration (y_{CO_2}, top), steam H_2O concentration (y_{H_2O}, mid) and flue gas volume flow ($\dot{V}$, bottom) at the calciner outlet.

6.2 Pilot plant operation

Stable plant operation has been achieved throughout all campaigns operating with a wide range of operation parameters. A time trend of one experiment aiming to demonstrate operation with low excess oxygen concentration is depicted in figure 6.3. This particular experiment was operated with a solid inventory of approx. $430 \, kg \, m^{-2}$ in the calciner and an average riser temperature of $933 \, °C$, whereas the carbonator held a solid inventory of approx. $680 \, kg \, m^{-2}$ with a carbonation temperature of $630 \, °C$. CO_2 capture was limited to approx. $0.80 \, mol \, mol^{-1}$ by the incoming amount of active CaO and averaged at $0.76 \, mol \, mol^{-1}$, eventually. Within the experiment the circulation rate was determined four times yielding consistent circulation rates of $390 \, kg \, h^{-1}$ or $1.35 \, kg \, s^{-1}$, respectively. The calciner was operated with an oxygen inlet concentration of $0.48 \, m^3 \, m^{-3}$ and an excess oxygen concentration of $0.02 \, m^3 \, m^{-3}$. On a dry and nitrogen free basis (as anticipated for commercial plants) the CO_2 purity of the flue gas stream averaged at $0.96 \, m^3 \, m^{-3}$.

Generally, operational errors resulting in temporary shut downs were mostly caused by malfunctions of the pilot plant's periphery such as the solid feeding system or the pressure control valves in the flue gas ducts. Minor fluctuations mostly associated with hydrodynamics were buffered by the system itself. Such a self-regulating effect can be seen after approx. 3.5 h. A rise in carbonation temperature with a simultaneous decrease in calcination temperature indicates that additional solids are moved from the calciner to the carbonator enhancing the CO_2 capture in the carbonator due to a surplus of CaO (second graph). The calciner operation with respect to flue gas quality or sorbent regeneration is hardly affected by the shift of solids as can be seen from the bottom graph of figure 6.3. Without active measures of the operator the system equalised to former operation. The shifting of solid inventory from the calciner to the carbonator and back to the calciner can also be observed by the reactors' solid inventory in the third graph of figure 6.3. The gradual decrease of the carbonator's solid inventory in this experiment results from excessive purging.

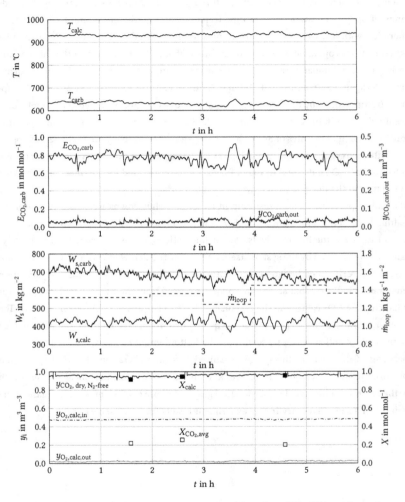

Figure 6.3: Time trend of an experiment operated for 6 h in CFB-CFB configuration.

6.3 Hydrodynamic operation

Stable hydrodynamic behaviour of the interconnected fluidised bed reactors is essential for the operation of a dual fluidised bed system such as calcium looping. Both, the BFB-CFB and the CFB-CFB configuration were used for the assessment of calcium looping CO_2 capture from cement plants. The BFB carbonator was primarily employed to assess make-up ratios that were out of the scope of the CFB-CFB configuration. However, the CFB carbonator is more likely to be employed for CO_2 capture from cement plants due to its superior flue gas to cross-section ratio as well as its enhanced gas-solid contact [26, 28]. An exemplary hydrodynamic profile of both pilot plant configurations is presented in figure 6.4. The pressure gradient of the reactors is depicted as solid lines, whereas internal circulation and transfer between the reactors are depicted as dashed lines and densely dotted lines, respectively. For both configurations, the pressure gradient of the calciner shows a smooth transition from the bottom section into a linear gradient at the calciner's riser section. In case of the BFB-configuration the mass in the calciner is increased to provide sufficient counter pressure to maintain the target mass in the BFB carbonator. Hence, a more dense region is formed at the reactor bottom (i.e. up to 1.5 m) before transitioning smoothly to the riser section. Also the pressure drop at the exit region is slightly more pronounced indicating initiating densification [79]. Due to the higher solid inventory more particles are reflected and internally recirculated resulting in an increased pressure drop at the riser exit region. However, densification phenomena are generally only observed for bench scale CFB reactors and are only expected for large scale CFB reactors if components are installed at the exit region restricting the gas-solid flow [26]. After separation in a cyclone the solids flow either into the standpipe of the conveyor screw or into the standpipe of the loop seal. Surplus solids that are not conveyed into the carbonator overflow into the calciner's loop seal. Due to space constraints during the construction of the pilot plant, only comparatively small standpipe heights above the loop seals can be maintained (i.e. pressure differences in the range of 30 mbar to 45 mbar) since the loop seals needed to be installed at a height of approx. 6 m to maintain an angle of approx. 60° for the reactors' return legs. Consequently, the pressure difference for internal recirculation fluctuates close to zero during operation. External circulation is less affected by this issue since solids exit the loop seal at the bottom of the stand pipe.

The CFB carbonator shows a similar pressure profile as the calciner. However, the riser section is less pronounced since solids are accumulated at the carbonator's bottom section due to its enlarged cross section. Approximately 80 % of the carbonator mass is accumulated

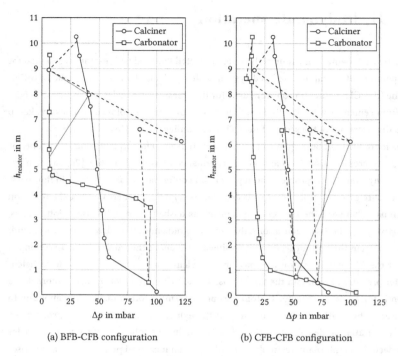

(a) BFB-CFB configuration   (b) CFB-CFB configuration

Figure 6.4: Hydrodynamic profiles depicted as reactor height (h_{reactor}) vs. pressure loss (Δp) of (a) the BFB-CFB configuration and (b) the CFB-CFB configuration investigating calcium looping CO_2 capture from cement plants.

in this section, whereas the remaining 20 % are attributed to the riser section. Solids in the transfer line (i.e. loop seals) are not included in this distribution but contribute to approx. 13 % of the overall mass of the carbonator. Furthermore, the pressure loss of the exit region as well as the cyclone is less developed due to lower gas velocities and lower particle density. The BFB carbonator shows a typical pressure gradient in which solids are accumulated at the reactor bottom and form a dense region with minor entrainment of solids into the freeboard region.

One major incident complicating the dual fluidised bed operation occurred when the CFB carbonator was operated with large CO_2 volume fractions representing low integration levels between the cement plant and the calcium looping process. The enlarged cross section at the bottom of the CFB carbonator was originally chosen to reduce the velocity of the flue gas to allow a longer residence time in the dense bed section of the carbonator. However, in case of high CO_2 flue gas concentrations a significant reduction of the flue gas volume flow occurred due to CO_2 capture. Consequently, the fluidisation gas velocity decreased resulting in insufficient solid entrainment from the dense region in the enlarged cross-section. In the course of the experiments adequate solid entrainment was ensured by increasing the total volume flow of flue gas fed to the carbonator maintaining the target CO_2 concentration. Thus, increasing the volume flow of CO_2 towards the carbonator and reducing the looping ratio as a consequence. Superficial velocity in the lower enlarged part of the reactor needed to be raised from $1.3\,\mathrm{m\,s^{-1}}$ up to $2\,\mathrm{m\,s^{-1}}$, corresponding to $3\,\mathrm{m\,s^{-1}}$ or $4.6\,\mathrm{m\,s^{-1}}$ in the carbonator's riser section (without CO_2 capture), to ensure proper solid entrainment and stable operation of the dual fluidised bed system. Depending on integration between the cement plant and the calcium looping unit the CO_2 flue gas concentration ranges up to $0.30\,\mathrm{m^3\,m^{-3}}$. Accordingly, the reduction of gas velocity can be as high as $30\,\%$. This reduction must and can easily be addressed by means of proper carbonator design. Despite the constraints of the pilot plant facility (i.e. low loop seal stand pipe heights enlarged bottom cross-section) stable operation over a wide range of operation parameters relevant for calcium looping CO_2 capture from cement plants have been successfully demonstrated indicating the applicability of the fluidised bed calcium looping technology for CO_2 capture from cement plants.

6.4 Carbonator CO_2 capture performance

The CO_2 capture in the carbonator is affected by various parameters such as the sorbent circulation rate, the sorbent capacity, the carbonation temperature and the residence time of the CO_2. These parameters are elaborated in the subsequent subsections. Eventually, the active space time approach is used to describe the CO_2 capture performance of the carbonator as a combination of these different variables, since the effect of the respective parameters superimposes each other.

6.4.1 Looping ratio and active looping ratio

The looping ratio describes the molar amount of CaO that is fed to the carbonator with respect to the molar amount of CO_2 entering the carbonator (eq. 2.6). Hence, it is a measure to determine the amount of CO_2 that can be removed from the gas phase by means of carbonation of CaO. Figure 6.5 presents the CO_2 capture efficiency for four different sorbent

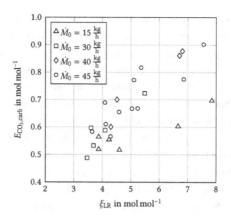

Figure 6.5: Carbonator CO_2 capture efficiency ($E_{CO_2,carb}$) vs. looping ratio (ξ_{LR}) for experiments investigating lower integration levels.

make-up rates, namely $15\,kg\,h^{-1}$, $30\,kg\,h^{-1}$, $40\,kg\,h^{-1}$ and $45\,kg\,h^{-1}$, investigating lower integration levels between the calcium looping process and the cement plant. The corresponding sorbent activity averaged around $0.112\,mol\,mol^{-1}$, $0.130\,mol\,mol^{-1}$, $0.129\,mol\,mol^{-1}$ and $0.135\,mol\,mol^{-1}$, respectively. In fact, the CO_2 carrying capacity of the $40\,kg\,h^{-1}$ sorbent make-up rate experiments is expected to exceed those of the $30\,kg\,h^{-1}$ experiments since the sorbent will undergo a reduced number of calcination and carbonation cycles with increasing make-up rate. The discrepancy can be explained by an interaction of various operational parameters affecting the actual number of calcination and carbonation cycles such as solid inventory and circulation rate as well as uncertainties in the determination of the CO_2 carrying capacity. For the presented experiments, CO_2 capture was neither limited by the calcination carbonation equilibrium nor the residence time but determined by the incoming amount of CaO. A linear increase of the CO_2 capture with the increasing looping ratio is evident. Since the CO_2 carrying capacity describes the absorbable amount of CO_2 per mole of CaO it defines the incremental improvement of the CO_2 capture efficiency with the looping ratio. Considering the uncertainties in sample collection and determination of the circulation rate, a good agreement between the determination of the inclination by linear regression and the determined CO_2 carrying capacity can be found. The respective results are summarised in table 6.1.

Table 6.1: Comparison between sorbent activity determined by linear regression and sorbent analysis.

Sorbent make-up $\dot{M}_0$ in kg h^{-1}	CO_2 carrying capacity $X_{CO_2,avg}$ in mol mol^{-1}	Inclination m in mol mol^{-1}	Agreement $X_{CO_2,avg}/m$
15	0.112	0.105	1.07
30	0.130	0.142	0.93
40	0.129	0.134	0.96
45	0.135	0.135	1.00

However, since only a fraction of the sorbent is active (i.e. able to absorb CO_2) the so called active looping ratio is applied to account for the sorbent's CO_2 carrying capacity (eq. 6.3). Per mole of active CaO one mole of CO_2 can be removed from the flue gas. Consequently, the active looping ratio represents a solid based CO_2 capture efficiency as long as no capture limitations such as insufficient reaction time or equilibrium constraints occur.

$$\xi_{LR,active} = \frac{\dot{N}_{CaO,loop}}{\dot{N}_{CO_2,carb,in}} \cdot X_{CO_2,avg} \tag{6.3}$$

In figure 6.6 the obtained CO_2 capture efficiency is plotted against the active looping ratio for all conducted experiments. According to their average CO_2 carrying capacity, the results are categorised into four categories. For an active looping ratio below 1 mol mol^{-1}, CO_2 capture in the carbonator is limited by insufficient amount of reactive CaO being fed into the carbonator. Consequently, the CO_2 capture increases linearly with increasing active looping ratio. The equilibrium mass balance is additionally depicted as solid line in figure 6.6. For active looping ratios above 1 mol mol^{-1}, surplus CaO is present in the carbonator and CO_2 capture efficiencies up to the equilibrium are achievable.

The experiments with an active looping ratio below 1 mol mol^{-1} show a good agreement with the postulated linear increase of the CO_2 capture efficiency with the active looping ratio. The deviation from the equilibrium balance averages at around 4.5 % and can again be mostly attributed to uncertainties in the determination of the solid circulation rate in combination with minor deviations between the reached sorbent loading and the determined sorbent CO_2 carrying capacity. As might be expected, the CO_2 carrying capacity of the experiments with a surplus of active looping ratio (i.e. above 1 mol mol^{-1}) was limited by equilibrium constraints. The experiments yielding comparatively low CO_2 capture efficiencies around 0.86 mol mol^{-1} for active looping ratios above 1 mol mol^{-1} correspond to the investigation of the calciner performance at various temperatures with a constant carbonation temperature of 700 °C. On a normalised basis, these CO_2 capture efficiencies correspond to

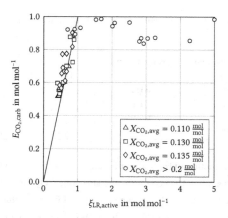

Figure 6.6: Carbonator CO_2 capture efficiency ($E_{CO_2,carb}$) vs. active looping ratio ($\xi_{LR,active}$) for all conducted fluidised bed experiments.

values close to 100 % as the equilibrium CO_2 capture averaged around $0.86 \, mol \, mol^{-1}$ for these experiments.

From these experiments it can be concluded that the CO_2 capture efficiency can be adjusted using the circulating amount of solids as long as the reaction time between the sorbent and the flue gas' CO_2 is sufficient and the equilibrium CO_2 partial pressure is not reached. Besides, the good agreement in the linear section indicated a good gas solid contact in the carbonator. A particular characteristic for the integration of calcium looping into a cement plant is that the cement plant's CO_2 emissions are linked to the sorbent make-up flow fed to the calcium looping system. Increasing the make-up flow concurrently reduces the amount of CO_2 fed to the carbonator since the CO_2 from sorbent calcination is not emitted in the cement plant's pre-calciner. Hence, for a given circulation rate the looping ratio and the sorbent activity increase if the make-up flow is increased. Consequently, the active looping ratio increases over-proportionately with increasing integration level making the make-up rate an effective lever to boost the CO_2 capture performance.

6.4.2   Carbonator temperature profile

The carbonator's temperature profile for the investigation of different carbonation temperatures with a sorbent make-up rate of $50 \, kg \, h^{-1}$ and a reference calcination temperature of approx. 918 °C is presented in figure 6.7. The temperature profiles are labelled accord-

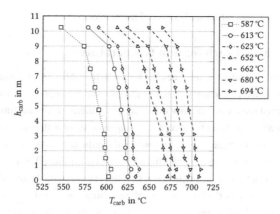

Figure 6.7: Carbonator temperature profiles for the investigation of different carbonation temperatures (T_{carb}) with a reference calcination temperature of approx. 918 °C.

ing to their reference carbonation temperature which was determined by integrating and averaging the temperature profile.

All temperature profiles run in parallel offset by the temperature in the dense region at the bottom of the carbonator. Opposite to the calciner's temperature profile, the carbonator's temperature remains constant for the first 1 m (i.e. the enlarged bottom section) were most of the solids are accumulated. In the carbonator's riser section the temperature decreases gradually. This decrease in temperature can be associated with a leaner particle concentration in the core annulus region reducing the heat released by carbonation as well as the thermal mass present in the carbonator riser. Hence, heat losses have a more pronounced impact on the temperature. The dense region's temperature is partly elevated due to the hot solid coming from the calciner entering the carbonator with a temperature of approx. 880 °C as well as due to the extensive heat release by the carbonation reaction. The heat supplied to the dense region by the carbonation reaction and the incoming solids from the calciner is removed by a bed heat exchanger resulting in the uniform temperature distribution in the dense region of the carbonator. The temperature in the dense bed region and implicitly in the riser section of the carbonator has been regulated by the heat exchanger's cooling duty.

The decreasing temperature profile of the carbonator affects the CO_2 capture in a beneficial way. In the hotter dense region, CO_2 uptake is enhanced due to improved carbonation kinetics, while simultaneously a low CO_2 partial pressure can be achieved in the exit region due to the reduced temperature in this part of the reactor.

6.4.3   Influence of carbonation temperature

The carbonation temperature determines the lowest achievable CO_2 concentration or partial pressure, respectively. Hence, it specifies directly the driving force of the carbonation reaction (i.e. partial pressure difference towards the equilibrium), the maximum amount of CO_2 that can be absorbed from the gas phase for a given volume flow of CO_2 and furthermore the carbonation reaction kinetics. However, the influence of the carbonation temperature is only detectable if no other limitations of the CO_2 capture efficiency such as insufficient sorbent circulation or insufficient reaction time are present. Figure 6.8 presents the CO_2 capture efficiencies for the investigation of high integration levels between the cement plant and the calcium looping process (i.e. high make-up rates and comparatively low CO_2 concentrations). For the presented experiments a surplus amount of active sorbent was circulated between the reactors. Consequently, CO_2 capture was not limited by the amount of circulating sorbent but determined by the carbonation temperature. The equilibrium CO_2 capture as a function of the carbonation temperature for a CO_2 concentration of $0.15\,\mathrm{m^3\,m^{-3}}$ is depicted as solid line. The experiments at a make-up ratio of $0.6\,\mathrm{mol\,mol^{-1}}$ conducted us-

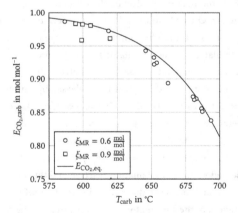

Figure 6.8: CO_2 capture efficiency of the carbonator ($E_{CO_2,carb}$) for various carbonation temperatures (T_{carb}) investigating high integration levels between the calcium looping CO_2 capture process and the cement plant (i.e. make-up ratios of $0.6\,\mathrm{mol\,mol^{-1}}$ and $0.9\,\mathrm{mol\,mol^{-1}}$ with a CO_2 concentration of $0.15\,\mathrm{m^3\,m^{-3}}$).

ing the CFB carbonator are depicted as circles and the experiments with a make-up ratio of $0.9\,mol\,mol^{-1}$ using the BFB carbonator are depicted as squares. The limitation of the CO_2 capture efficiencies by the equilibrium is evident. The CO_2 capture increases from $0.84\,mol\,mol^{-1}$ at a carbonation temperature of 690 °C up to $0.98\,mol\,mol^{-1}$ at temperatures slightly below 600 °C. All experiments yield a normalised CO_2 capture efficiency in the range of 95 % to 99 %.

Generally, when operating with highly cycled sorbent it was concluded that the optimal carbonation temperature lies around 650 °C resulting from a trade-off between the achievable CO_2 partial pressure and the carbonation reaction rate [29, 74, 156]. However, when operating with low cycled sorbent, which holds a significantly higher surface area, temperatures around 600 °C proved to be beneficial with respect to CO_2 capture due to the lower achievable CO_2 concentration. These results are in agreement with the findings of Lu et al. [116] who obtained the best CO_2 capture performance at a carbonation temperature range of 580 °C to 600 °C for low cycled sorbent due to its increased porosity.

6.4.4   Sorbent CO_2 carrying capacity

For constant calcination conditions, the sorbent CO_2 carrying capacity is primarily affected by the number of calcination and carbonation cycles. With increasing make-up rate the sorbent's residence time in the calcium looping system decreases and the sorbent activity increases. In case of calcium looping CO_2 capture from cement plants the sorbent make-up rate as well as the CO_2 flue gas load towards the carbonation reactor is defined by the integration level between the cement plant and the calcium looping unit. Figure 6.9 presents the obtained average CO_2 carrying capacity ($X_{CO_2,avg}$) as a function of the number of calcination and carbonation cycles (n_{cycle}) for all conducted fluidised bed experiments. The number of calcination and carbonation cycles was determined using the product of the cycle frequency (i.e. first fraction of eq. 6.4) and the sorbent's residence time in the calcium looping system (i.e. second fraction of eq. 6.4).

$$n_{cycle} = \frac{\dot{N}_{CaO,loop}}{N_{CaO,carb} + N_{CaO,calc}} \cdot \frac{N_{CaO,carb} + N_{CaO,calc}}{\dot{N}_{CaCO_3,0}} \tag{6.4}$$

Generally, the experiments screening low integration levels (i.e. reduced make-up rates and high CO_2 concentrations) yielded average CO_2 carrying capacities between $0.10\,mol\,mol^{-1}$ and $0.16\,mol\,mol^{-1}$ depending on the make-up rate. Whereas, the experiments investigating high integration levels yielded sorbent activities above $0.2\,mol\,mol^{-1}$. Figure 6.9 further

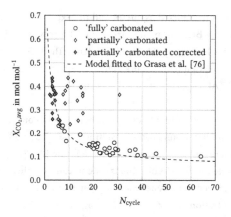

Figure 6.9: Sorbent CO_2 carrying capacity ($X_{CO_2,avg}$) vs. number of calcination and carbonation cycles (n_{cycle}) for all experiments investigating fluidised bed calcium looping CO_2
capture from cement plants.

comprises the model line of the utilised limestone fitted to the deactivation model proposed
by Grasa et al. [76]. Two groups of data can be distinguished. Samples that show a fairly
good agreement between the predicted and the measured sorbent activity (white circles)
and samples that show a larger deviation between the predicted and measured CO_2 carrying capacity (white diamonds). It is evident that especially samples with a lower sorbent
activity match the model fairly well, whereas in many cases solid samples holding a high
sorbent activity deviate severely. The high deviation of these samples can be ascribed to
partial carbonation of the circulating solids. In these experiments the surplus amount of active CaO present in the carbonator could not be fully carbonised (i.e. reach its CO_2 carrying
capacity) either due to insufficient CO_2 being fed to the carbonator or due to equilibrium
constraints imposed by the carbonation temperature. This phenomenon occurred for high
make-up rates of fresh sorbent once the BFB carbonator was employed as well as at higher
carbonation temperatures if the CFB carbonator was used with high sorbent make-up rates.
At higher carbonation temperatures sorbent carbonation was restricted due to the achievable CO_2 concentration or the absorbable amount of CO_2, respectively. Whereas, in case
of the BFB carbonator insufficient amount of CO_2 was fed to the carbonator to reach the
sorbent's CO_2 carrying capacity.
Partially carbonated samples undergo only a reduced effective number of calcination and
carbonation cycles resulting in a younger sorbent age [137]. Hence, samples with fractional

carbonation degree below $0.5\,\mathrm{mol\,mol^{-1}}$ were corrected using equation 6.6 assuming that the conducted number of calcination and carbonation cycles is proportional to the sorbent's fractional carbonation conversion (eq. 6.5). The term fractional carbonation (f_{carb}) describes the amount of CO_2 absorbed ($X_{\mathrm{CaCO_3,carb}} - X_{\mathrm{CaCO_3,calc}}$) with respect to the sorbent's CO_2 carrying capacity ($X_{\mathrm{CO_2,avg}}$).

$$f_{\mathrm{carb}} = \frac{X_{\mathrm{CaCO_3,carb}} - X_{\mathrm{CaCO_3,calc}}}{X_{\mathrm{CO_2,avg}}} \tag{6.5}$$

$$n_{\mathrm{age}} = n_{\mathrm{cycle}} \cdot f_{\mathrm{Carb}} \tag{6.6}$$

Correcting the partially carbonated sample according to their cycle age (grey diamonds) yield a fairly good agreement with the predicted sorbent CO_2 carrying capacity. Unsurprisingly, sorbent samples with a low fractional carbonation conversion coincide with experiments operated with a surplus amount of active looping ratio (i.e. above $1.5\,\mathrm{mol\,mol^{-1}}$). Nonetheless, it has to be highlighted that a state in which excess sorbent is circulated between the reactor that does not participate in CO_2 capture represents an inefficient state of operation. The non-reacting sorbent circulated between the reactors will cause needless heat losses and requires over-sizing of the calcium looping plant.

6.4.5 Influence of flue gas CO_2 concentration

Higher CO_2 partial pressures increase the carbonation reaction's driving force, thus, accelerating the carbonation reaction. However, as CO_2 capture progresses the CO_2 partial pressure reduces towards the equilibrium CO_2 partial pressure and thereby, reducing the carbonation reaction rate gradually. From a CO_2 capture perspective the initially accelerated carbonation reaction plays a subordinate role since the proportion of active CaO and CO_2 defines the carbonator's CO_2 capture potential as long as sufficient reaction time is provided by means of reactive solid carbonator inventory. Throughout the conducted experiments the CO_2 capture performance was hardly affected by the flue gas' CO_2 concentration but by the amount of CaO able to react with the CO_2 since all experiments were operated with excess reaction time. Hence, a positive influence of the higher CO_2 partial pressures accelerating the carbonation reaction could not be determined in this work.

6.4.6 Carbonator active space time approach

The active space time approach developed by Charitos et al. [37, 39] represents a simplified carbonator CO_2 balance that can be used for a basic design of the calcium looping

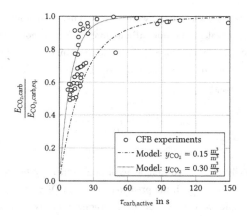

Figure 6.10: Normalised CO_2 capture efficiency ($\frac{E_{CO_2,carb}}{E_{CO_2,carb,eq.}}$) vs. active space time ($\tau_{carb,active}$) for experiments conducted with the CFB carbonator investigating high and low integration levels between the calcium looping CO_2 capture process and the cement clinker production. Model lines calculated for a carbonation temperature of 650 °C.

carbonator (section 2.4.1). The term active space time ($\tau_{carb,active}$) combines the parameters adjustable by means of process design or operation such as sorbent activity (by sorbent make-up), circulation rate and solid carbonator inventory.

Figure 6.10 presents the normalised CO_2 capture efficiency against the active space time for all conducted CFB-CFB experiments. To account for variations of the carbonation temperatures the normalised CO_2 capture efficiency is chosen as ordinate. The normalised CO_2 capture efficiency for a carbonation temperature of 650 °C predicted by the active space time model for a CO_2 concentration of $0.15\,\mathrm{m^3\,m^{-3}}$ and $0.30\,\mathrm{m^3\,m^{-3}}$ are depicted as dotted and densely dotted lines, respectively. Those model lines represent the lower and upper limit of CO_2 concentrations anticipated for calcium looping CO_2 capture from cement plants with high and low integration levels. A good agr eement between the conducted experiments and the active space time model is evident as all CO_2 capture efficiencies lie mostly between the predicted model lines.

By means of TGA analysis of the collected solid samples, the sorbent's specific reaction constant (k_{carb}), the sorbent average CO_2 carrying capacity ($X_{CO_2,avg}$) and critical carbonation time (t^*_{carb}) were determined. Furthermore, chemical analyses were conducted to assess the elemental composition of the carbonator and calciner solids. Process parameters, such as the circulation rate ($\dot{M}_{loop}$, $\dot{N}_{loop,CaO}$), the carbonator inventory (M_{carb}, $N_{Ca,carb}$), the molar flow of CO_2 entering the carbonator ($\dot{N}_{CO_2,carb}$) and the CO_2 concentrations were derived from data recorded during the experiments in combination with the chemical analysis of the collected solid samples. The reaction rate constant of the utilised limestone was determined to be $0.3393\,s^{-1}$ with a standard deviation of $0.0560\,s^{-1}$ averaging all conducted TGA analysis. The determined reaction rate constant matches those reported in the literature quite well. Reported limestone reaction constants range from $0.20\,s^{-1}$ to $0.37\,s^{-1}$ [39, 138]. Fitting the experimentally obtained CO_2 capture efficiencies to the determined active space times yield an apparent reaction rate ($k_{carb}\varphi$) of $0.30\,s^{-1}$ with a corresponding gas solid contact factor (φ) of 0.92 for the CFB experiments, whereas the BFB experiments yield an apparent reaction rate of $0.09\,s^{-1}$ with a gas solid contact factor (φ) of 0.24. The distinctive poor gas solid contact of the BFB carbonator results from bubble formation allowing the flue gas to by-pass the sorbent. Contrary, the CFB carbonator shows very good gas solid contact due to the operation in the fast fluidisation regime.

From the experimental results is can be concluded that an active space time of $50\,s$ is sufficient to achieve a normalised CO_2 capture efficiency of $90\,\%$ corresponding to an overall carbonator CO_2 capture efficiency of $84\,\%$ for a carbonation temperature of $650\,°C$ and a flue gas CO_2 concentration of $0.15\,m^3\,m^{-3}$. With increasing CO_2 concentration or decreasing integration level, respectively, the required active space time is reduced as the driving force of the carbonation reaction increases. Nonetheless, the parameters determining the active space time are significantly affected by the integration level. With decreasing integration level the make-up rate decreases and therefore, the sorbent activity decreases while simultaneously the amount of CO_2 fed to the carbonator increases. Thus, providing a certain active space time to achieve a target CO_2 capture efficiency will demand more efforts in terms of circulation rate (i.e. active particle fraction) or solid inventory (i.e. residence time).

6.5   Calciner operation

The main objectives of the calcium looping calciner are (i) the regeneration of the sorbent to enable repetitive CO_2 capture in the carbonator and (ii) the supply of a CO_2 enriched flue gas stream suitable for storage or utilisation after purification in a CPU. The CPU performance depends primarily on the CO_2 concentration or, indirectly, on the amount of impurities such as oxygen or nitrogen being present in the treated flue gas. Soluble impurities such as NO_x and SO_2 can easily be removed during the compression steps of the CPU [154, 179]. The sorbent performance is mainly affected by the conducted calcination and carbonation cycles but also depends on the calcination conditions. Sorbent deactivation is enhanced at elevated calcination temperatures and high CO_2 partial pressures as those conditions promote severe sintering. Hence, calcination temperatures should be chosen as low as possible while ensuring sufficiently fast sorbent calcination to maintain a desired sorbent regeneration. In the following subsections the calciner's regeneration performance as well as the flue gas quality will be elaborated.

6.5.1   Calciner temperature profile

The temperature profiles of the conducted variations of the calcination temperature at a constant make-up flow of $50\,\mathrm{kg\,h^{-1}}$ are presented in figure 6.11, whereas the variations of sorbent make-up at a constant calcination temperature of approx. $920\,°C$ are depicted in figure 6.12. The calciner's temperature profile is characterised by a gradual increase from the bottom of the reactor to approx. $4\,\mathrm{m}$ or the last stage of oxidant feeding, respectively. The gradual increase of the calciner temperature can be attributed to heat sinks entering the calciner at the reactor bottom and the fact that fuel combustion is incomplete due to under stoichiometric operation conditions up to $4\,\mathrm{m}$. Once all oxidant is fed to the calciner the combustion conversion completes and the temperature remains constant. As a consequence, the average temperature from $4\,\mathrm{m}$ to the reactor exit is chosen as reference temperature for the calciner. The temperature profile can be ascribed to three major heat control mechanisms namely (i) endothermic calcination reaction, (ii) oxidant staging and (iii) sensible heat from flue gas recirculation and solid circulation. The calcination of circulating sorbent and sorbent make-up flow (i.e. $CaCO_3$) forms a major heat sink due to the endothermic nature of the calcination reaction. Initially, the sorbent is continuously heated to its calcination temperature. Once the calcination temperature is reached, the particles remain at a steady-state temperature until calcination is completed. This 'self-cooling' effect of the calcining particles smooths the temperature profile of the calciner. The recircu-

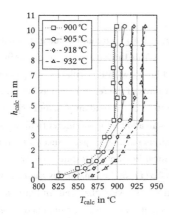
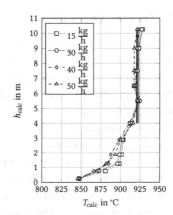

Figure 6.11: Calciner temperature profile for different calcination temperatures at a sorbent make-up rate of 50 kg h^{-1}.

Figure 6.12: Calciner temperature profile for different sorbent make-up rates at a calcination temperature of approx. 920 °C.

lated flue gas represents a sensible heat sink similar to the circulating solids. However, the amount of recirculated flue gas can be chosen independently from other process parameters bearing in mind that a minimal fluidisation gas flow needs to be maintained. Consequently, the amount of recirculated flue gas can be used to directly influence the calciner's temperature profile. Besides, staged oxidant feeding enables a more precise and more uniform release of the fuel's combustion heat preventing the formation of local hotspots especially close to the fuel feeding point (i.e. lower part of the calciner). These temperature control mechanisms in combination with the good heat transfer and heat distribution properties of CFB reactors lead to the uniform temperature profile without severe hotspots observed for all experiments. Such operation conditions are beneficial for sorbent calcination regarding its CO_2 capture performance since sintering (i.e. sorbent deactivation) is minimised. It is apparent from figure 6.11 and figure 6.12 that the above described temperature profile have been observed regardless of the operation temperature or the sorbent make-up feed. Thus, emphasising that such beneficial sorbent regeneration conditions are achievable independent from calcination temperature or sorbent make-up rate using CFB reactors with staged oxidant feeding.

6.5.2   Oxy-fuel operation

In case of oxy-fuel combustion, pure oxygen is diluted with recirculated flue gas (mostly CO_2) in order to control the combustion temperature. While the fed amount of fuel defines the required oxygen to fully combust the fuel, the dilution with recirculated flue gas can be independently adjusted maintaining certain constraints of the reactor such as minimum fluidisation. As a consequence, the oxygen concentration of the oxidant gas, more specifically the fluidisation gas or superficial volume flow, is an additional parameter to influence and control the fuel combustion in the calciner. With increasing oxygen concentration or decreasing dilution the system is prone to exhibit local hotspots. Despite operating the calciner with an oxygen inlet concentration of up to $0.57\,\mathrm{m^3\,m^{-3}}$ no major deviations from the temperature profile previously described in section 6.5.1 were observed. In figure 6.13 the oxygen concentration is plotted against the recirculation rate for all conducted experiments. Additionally, selected temperature profiles for various oxygen concentrations are depicted in figure 6.14. The respective experiments are highlighted as black marks in figure 6.14.

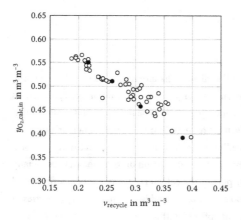

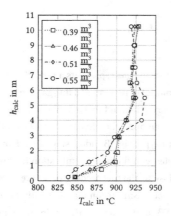

Figure 6.13: Inlet oxygen concentration on a wet basis ($y_{O_2,calc,in}$) vs. flue gas recirculation rate ($v_{recycle}$) for all conducted fluidised bed experiments.

Figure 6.14: Selected calciner temperature profiles for different inlet oxygen concentrations at a calcination temperature of approx. 920 °C

The selected temperature profiles are characterised by a superficial velocity of approx. $3.1 \, \text{m s}^{-1}$ (at a reference height of 4 m, once all fluidisation gas is fed to the calciner), a solid inventory of approx. 20.8 kg (or $440 \, \text{kg m}^{-2}$) and a reference calcination temperature of 920 °C. As can be seen in figure 6.14 the temperature profiles are hardly influenced by the increasing oxygen inlet concentration. Thus, beneficial sorbent calcination conditions can be maintained at high inlet oxygen concentration allowing to reduce the amount of recirculated flue gas. However, the $0.55 \, \text{m}^3 \, \text{m}^{-3}$ experiment shows a slightly increased temperature up to around 936 °C (approx. 10 K) above 4 m once the last fraction of oxidant gas is fed to the calciner. Nonetheless, such minor increase in temperature will hardly cause enhanced sorbent sintering. The comparatively small influence of flue gas recirculation on the calciner's temperature can be explained by the more prominent influence of sorbent calcination and the sorbent heat up. Throughout the experiments, the energy required for sorbent calcination (circulating and make-up) ranged from 40 kW to 80 kW corresponding to approx. 50 % of the calciner's energy duty. The heat up of the circulating solids (sensible heat) accounted for 10 kW to 40 kW, whereas the recirculated flue gas accounts for only approx. 7 kW to 10 kW or between 5 % and 10 % of the calciner's energy duty, respectively.

From these results it can be concluded that the additional energy sinks available in the calcium looping calciner – especially the calcination of captured CO_2 (i.e. circulating $CaCO_3$) and sorbent make-up – facilitate the reduction of the flue gas recirculation rate while simultaneously maintaining a smooth and uniform temperature profile. Even operation without recirculation is conceivable if sufficient fluidisation gas is initially provided by oxygen feeding and subsequently by combustion and sorbent calcination. For such operation staged fuel feeding might be beneficial to allow a dedicated heat release by staged combustion as oxygen would be required as initial fluidisation agent. The operation with reduced flue gas recirculation rates or even without any recirculation will reduce the size of the recycle line and equipment and eventually operation and investment costs of the calcium looping unit.

6.5.3   Calciner active space time approach

In this section the calciner's sorbent regeneration performance is assessed using the calcination efficiency (E_{calc}) as well as the active space time approach adapted to the calcination reaction. In addition, the approach of Martinez et al. [124] has been slightly modified to address the large share of sorbent make-up on the $CaCO_3$ load for instance in case of high integration levels. Due to the large discrepancy between the fresh sorbent's and the circulated sorbent's carbonate content their respective active particle fraction ($f_{active,0}$, $f_{active,loop}$)

has been calculated individually using equations 2.28 to 2.31. The effective active particle fraction has subsequently been determined by superposition of the two particle fraction's respective proportion (eq. 6.8). Modifying the approach of Martinez et al. [124] as described, the modified approach converts towards the original approach for low shares of sorbent make-up (ξ_0, eq. 6.7).

$$\xi_0 = \frac{\dot{N}_{CaCO_3,0}}{\dot{N}_{CaCO_3,0} + \dot{N}_{CaCO_3,loop}} \tag{6.7}$$

$$f_{active,eff} = f_{active,0} \cdot \xi_0 + f_{active,loop} \cdot (1 - \xi_0) \tag{6.8}$$

Throughout the experiments the circulating solids (partially carbonated) required around 3 s to achieve full calcination, whereas the fully carbonated sorbent make-up required approx. 20 s. For the conducted experiments, the share of make-up on the carbonate content entering the calciner (ξ_0) ranged from $0.2\,mol\,mol^{-1}$ at low integration levels to up to $0.5\,mol\,mol^{-1}$ at high integration levels.

The obtained regenerator efficiencies are presented in figure 6.15 against the active space time evaluated according to the approach by Martinez et al. [124] (i.e. average carbonate content), whereas figure 6.16 presents the regenerator efficiency against the active space time calculated using superposition of the particular active particle fractions (i.e. modified approach). Additionally, the model lines for a calcination temperature of 900 °C, 918 °C and 930 °C at a CO_2 partial pressure of 0.49 bar, representing the logarithmic average CO_2 partial pressure of the experiments, are depicted. In case of the modified approach the dotted lines represent a sorbent make up share of $0.2\,mol\,mol^{-1}$ on the carbonate flow entering the calciner as obtained during the screening of lower integration levels. The dashed line refers to a share of $0.4\,mol\,mol^{-1}$ corresponding to the high integration experiments at a calcination temperature of 918 °C.

The determined calcination efficiencies follow the limited growth pattern predicted by the active space time model fairly well. In the investigated temperature range, no major dependency of the calcination temperature on the calcination efficiency was observed. Opposite to this observation the model predicts an increase of the regeneration efficiency with increasing calcination temperature. The increase diminishes at higher calcination temperatures since faster reaction kinetics increase the calcination reaction and simultaneously decrease the active particle fraction. The discrepancy between the experiments and the model predictions may result from uncertainties in the determination of the parameters required for the calculation of the active particle fraction such as the circulation rate or the sorbent carbonation degree and eventually, the active space time. Furthermore, the investigated temperature range is quite narrow, impeding an unambiguous assessment concerning

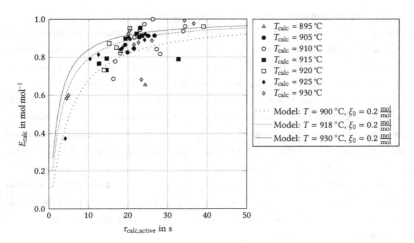

Figure 6.15: Calciner regeneration efficiency (E_{calc}) vs. the calciner active space time ($\tau_{calc,active}$) calculated as premixed flow of the circulating $CaCO_3$ and sorbent make-up fraction.

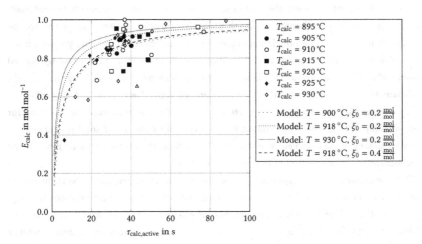

Figure 6.16: Calciner regeneration efficiency (E_{calc}) vs. the calciner active space time ($\tau_{calc,active}$) calculated by superposition of the circulating $CaCO_3$ and sorbent make-up fraction.

the effect of calcination temperature. Nonetheless, sufficient agreement of the experiments with the predicted model confirms the model's applicability. Comparing the two graphs, it is evident that both approaches can adequately describe the obtained regenerator efficiencies. It is noteworthy that the standard approach yields significantly lower active space times. This is due to the fact that the active particle fraction is strongly reduced since the the mixed carbonate content is used for the calculation of the active particle fraction. Still, both active space times correspond to the same particle residence time. With increasing share of sorbent make-up the required active space time to achieve a target regeneration efficiency increases. This effect is most prominent for high regenerator efficiencies as the sorbent make-up's prolonged time required to achieve full calcination becomes more dominant. According to the modified approach at a calcination temperature of 918 °C an active space time of 32 s is required for a sorbent make up share of 0.2 mol mol^{-1} to achieve a regeneration efficiency of 0.9 mol mol^{-1}, whereas approx. 50 s of active space time are required for a make-up share of 0.4 mol mol^{-1}.

Despite the large difference in required calcination time of the two particle fractions, both models adequately predict the calciner's calcination efficiency. Since the reference approach over-predicts the active fraction of circulating sorbent and under-predicts the active fraction of sorbent make-up, the overall active particle fraction is somewhat counterbalanced. In case of the reference approach the active particle fraction averaged at around 0.025. Applying the superposition approach to all conducted experiments yielded an averaged active particle fraction of 0.042. The active fraction of sorbent make-up ranged from 0.086 at high integration levels to 0.135 for the screening of lower integration levels, whereas the circulating sorbent yielded an averaged active fraction of 0.017 regardless of the make-up rate. The larger active particle fraction of the sorbent make-up observed for the screening of the lower integration levels is caused by the lower solid residence time in the calciner as more sorbent was circulated to address the higher CO_2 load towards the carbonator in combination with the lower sorbent activity.

While the active space time approach can be used to describe the regeneration performance of the calciner the immediate use is less intuitive compared to the active space time approach of the carbonator, due to the fact that the calcination reaction kinetics and the carbonate content directly influence the active particle fraction by the required time to achieve full calcination. The carbonate content further affects the calciner's active space time as the carbonates' residence time is used as a basis for the calculation. Further on, the calcination kinetics directly affects the regeneration efficiency for a specific active space time (eq. 2.30). It has to be highlighted that the calculated active space times of both approaches correspond to the same particle residence time. Assuming a CSTR behaviour of the solid phase, the residence time of the two particle fractions cannot be distinguished as they are mixed promptly. Which approach is most suitable to describe the calcination performance cannot be conclusively determined. The application of both approaches as basis for the design of the calcination reactor of a calcium looping system is reasonable. The modified approach will more adequately describe the calcination performance in case of high make-up to circulating carbonate ratios due to the fact that the different reaction times to achieve full calcination are taken into account. Then again, a perfectly mixed solid phase anticipated for CFB reactors supports the utilisation of an average carbonated content to represent all solids.

6.6 Calciner flue gas composition

A major criterion regarding subsequent storage or utilisation of the CO_2 rich calciner flue gas stream is its composition, most importantly the CO_2 concentration, but also impurities such as excess oxygen, nitrogen oxides or sulphur oxides. While nitrogen and oxygen increase the CPU energy consumption due to a heightened separation duty, NO_x and SO_2 may cause corrosion due to the formation of nitric acid or sulphuric acid, respectively. Figure 6.17 presents selected gas compositions for various calcination temperatures with a constant sorbent make up flow of 50 kg h^{-1} and for a constant calcination temperature of 920 °C with four make-up rates namely 15 kg h^{-1}, 30 kg h^{-1}, 40 kg h^{-1} and 50 kg h^{-1}. Throughout the conducted experiments only minor deviations in the gas composition occurred. The most prominent influence originates from N_2 dilution or more specifically the total flue gas volume flow. Certain units of the pilot plant need to be flushed or operated with nitrogen, hence, the calciner's flue gas is diluted with nitrogen to a certain extent. As the nitrogen purge flow stays rather constant, the nitrogen dilution is more prominent at lower flue gas volume flows. The nitrogen concentration ranged from 0.10 m^3 m^{-3} to 0.14 m^3 m^{-3}. Overall, the CO_2 concentration averaged at around 0.66 m^3 m^{-3}, the H_2O concentration at around 0.19 m^3 m^{-3}, whereas the oxygen concentration ranged from 0.02 m^3 m^{-3} to 0.06 m^3 m^{-3}. On a dry and nitrogen free-basis the CO_2 concentration averaged at around 0.95 m^3 m^{-3}. Such a state can be anticipated for a full scale calcium looping plant after flue gas cooling below the dew point (i.e. before the CPU).

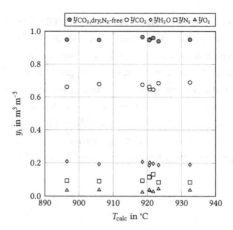

Figure 6.17: Calciner flue gas composition (y_i) for various calcination temperatures (T_{calc}) and a constant make-up flow of $50\,kg\,h^{-1}$ as well as for various sorbent make-up flows ($\dot{M}_{CaCO_3}$) at the reference calcination temperature of 920 °C.

6.6.1   CO$_2$ emissions

The CO$_2$ purity of the calciner's flue gas was reduced due to a dilution with nitrogen since the loop seal needed to be operated with nitrogen in order to avoid carbonation and consequently defluidisation of solids in the loop seal as temperatures slightly below 900 °C usually occurred. Furthermore, nitrogen is used to purge the pressure transducer lines and the driving shaft of the recirculation gas blower as well as the rotary valve. Throughout the conducted experiments the CO$_2$ concentration ranged from $0.60\,m^3\,m^{-3}$ to $0.70\,m^3\,m^{-3}$ averaging at $0.66\,m^3\,m^{-3}$. The origin of the flue gas' CO$_2$ is presented in figure 6.18 against the flue gas recirculation ratio. Additionally, the calciner flue gas' CO$_2$ concentration is depicted as measured and as anticipated for a full scale calcium looping plant before the CPU (i.e. on a dry and nitrogen free basis).

Due to comparatively high heat losses of the pilot facility the amount of fuel burnt and hence, the fraction of CO$_2$ associated to fuel combustion stayed rather constant averaging at around $0.41\,m^3\,m^{-3}$ or $35\,m^3\,h^{-1}$. Naturally, the share of recirculated flue gas on the overall CO$_2$ decreased with decreasing recirculation ratio while the share of CO$_2$ related to sorbent calcination (i.e. capture CO$_2$ and sorbent make-up) increased accordingly.

From figure 6.18 it becomes evident that the operation conditions in the calciner (sorbent

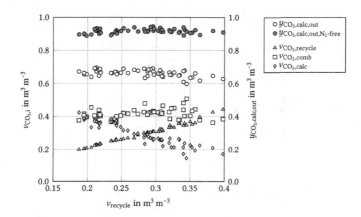

Figure 6.18: Share of fuel combustion, sorbent calcination and recirculated flue gas on flue gas CO_2 ($v_{CO_2,i}$) vs. the recirculation ratio ($v_{recycle}$). Additionally, the flue gas CO_2 concentrations at the calciner outlet as measured ($y_{CO_2,calc,out}$) and as anticipated for commercial scale ($y_{CO_2,calc,out,N_2-free}$) are depicted.

make-up rate, sorbent circulation rate, calcination temperature, O_2 inlet concentration, ...) hardly affect the obtained CO_2 purity, excluding excess oxygen. Hence, regardless of operation or boundary conditions a CO_2 concentration around $0.95\,m^3\,m^{-3}$ can be anticipated entering the CPU of a commercial calcium looping system. Hereby, the CO_2 purification is facilitated resulting in reduced size and energy consumption of the CPU [107, 153].

6.6.2 Excess O_2 and CO emissions

In case of oxy-fuel operation, oxygen supply is decoupled from the fluidisation gas of the CFB calciner. Hence, the amount of excess oxygen can directly be controlled by the oxygen supply, considering that a certain excess oxygen must be foreseen to ensure a satisfying fuel burnout. Generally, CO emission decreased with increasing excess oxygen and increasing combustion temperature (figure 6.19). With decreasing combustion temperature, the enhancing effect of excess oxygen on CO emission weakens. For an excess oxygen concentration of $0.04\,m^3\,m^{-3}$ the CO emission increased from $20\cdot10^{-6}\,m^3\,m^{-3}$ for temperatures above $920\,°C$ to $40\cdot10^{-6}\,m^3\,m^{-3}$ for temperatures between $920\,°C$ and $910\,°C$ and further increased to $50\cdot10^{-6}\,m^3\,m^{-3}$ for temperatures below $910\,°C$. For temperatures below $910\,°C$ the reduction of CO emission with increasing excess oxygen was limited. An increase of ex-

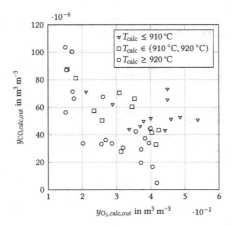

Figure 6.19: CO emissions of the oxy-fuel calciner ($y_{CO,calc,out}$) vs. excess oxygen concentration ($y_{O_2,calc,out}$) for all conducted fluidised bed experiments.

cess oxygen concentration above $0.04\,\mathrm{m^3\,m^{-3}}$ did not result in a further reduction of the CO emissions. These findings are in agreement with those reported for oxy-fuel CFB combustion [83, 103, 108, 126]. Furthermore, Hofbauer [83] reported decreasing CO emission with increasing temperature and found CO emissions around $200 \cdot 10^{-6}\,\mathrm{m^3\,m^{-3}}$ at a combustion temperature of $900\,°C$ as well as a negligible influence of excess oxygen on CO emission above $0.02\,\mathrm{m^3\,m^{-3}}$ in the investigated temperature range of $850\,°C$ to $910\,°C$. It has to be highlighted that the CO emissions presented here are reduced compared to those reported by Hofbauer as the flue gas is diluted with CO_2 from sorbent calcination compared to pure oxy-fuel combustion. The share of CO_2 originating from sorbent calcination accounted for $10\,\%$ to $25\,\%$ of the total volume flow.

Based on these results, it can be concluded that the dependency of CO emissions in case of calcium looping operation is consistent with those reported for pure oxy-fuel combustion.

6.6.3   SO$_2$ emissions

Generally, a caustic scrubber is foreseen subsequent to the oxy-fuel combustion or rather before the CPU unit to remove acid gas components such as SO_2, HCl or HF and to avoid the formation of their respective acids during compression [115]. Figure 6.20 presents the measured SO_2 concentration of the calciner flue gas, whereas their corresponding SO_2 cap-

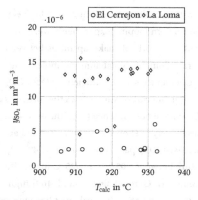

Figure 6.20: SO$_2$ emissions (y_{SO_2}) vs. calcination temperature (T_{calc}) for the two burnt Colombian hard coals.

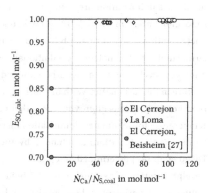

Figure 6.21: SO$_2$ capture efficiency ($E_{SO_2,calc}$) vs. molar calcium to sulphur ratio ($\frac{\dot{N}_{Ca}}{\dot{N}_{S,coal}}$)

ture efficiencies are presented in figure 6.21. Throughout the conducted experiments low SO$_2$ concentrations averaging around $2 \cdot 10^{-6}$ m^3 m^{-3} were measured when burning the El Cerrejon coal with a sorbent make-up of 50 kg h^{-1} and around $13 \cdot 10^{-6}$ m^3 m^{-3} when utilising the La Loma coal with a sorbent make-up of 45 kg h^{-1}. Due to a damage at the gas analyser's SO$_2$ sensor, no SO$_2$ data is available for lower sorbent make-up rates. The presented operation conditions correspond to molar calcium to sulphur ratios of around 100 mol mol^{-1} and 50 mol mol^{-1}, respectively. As expected, high SO$_2$ capture efficiencies of 0.992 mol mol^{-1} and 0.998 mol mol^{-1} were reached due to the high amount of calcium available for desulphurisation. The obtained SO$_2$ emissions are significantly reduced compared to oxy-fuel combustion experiments conducted at University of Stuttgart's fluidised bed pilot facility by Hofbauer [83] and Beisheim [27] burning the same Colombian hard coal (El Cerrejon). Hofbauer [83] measured SO$_2$ concentrations in the range of $1200 \cdot 10^{-6}$ m^3 m^{-3} to $1900 \cdot 10^{-6}$ m^3 m^{-3} without in-situ desulphurisation, whereas Beisheim [27] measured SO$_2$ concentrations between $300 \cdot 10^{-6}$ m^3 m^{-3} and $600 \cdot 10^{-6}$ m^3 m^{-3} during in-situ desulphurisation experiments operated with a calcium to sulphur ratio of 3 mol mol^{-1} at temperatures of 842 °C, 875 °C and 916 °C.

The SO$_2$ concentrations observed during calcium looping operation were hardly affected by the operation conditions. Partly because of the operation in a rather narrow calcination temperature range as well as an exceptionally high supply of calcium. CFB boilers with in-situ desulphurisation are generally operated with a calcium to sulphur ratio between $2\,\mathrm{mol\,mol^{-1}}$ and $3\,\mathrm{mol\,mol^{-1}}$ [18, 26], whereas the calcium looping experiments were conducted with a calcium to sulphur ratio associated with sorbent make-up ratios between $20\,\mathrm{mol\,mol^{-1}}$ to $100\,\mathrm{mol\,mol^{-1}}$. Additionally, several kilomoles of calcium per mole of sulphur were cycled between the carbonator and the calciner to provide sufficient CaO for CO_2 capture in the carbonator. With increasing calcination and carbonation cycles the sorbent's pore network widens improving the sorbent's SO_2 uptake performance [77]. The optimum temperature for in-situ desulphurisation increases for oxy-fuel operation since sorbent activation (i.e. calcination) is inhibited by the increased CO_2 partial pressure [27, 49, 69]. Diego et al. [49] determined an optimal temperature between 900 °C and 925 °C for in-situ desulphurisation of oxy-fuel fired CFB boilers. Similarly, García-Labiano et al. [69] highlighted the superiority of desulphurisation by indirect sulphation (i.e. calcined limestone) over the direct sulphation pathway, yielding the best desulphurisation performance at temperatures between 900 °C and 925 °C. For higher temperatures SO_2 conversion decreased due to enhanced sintering [69]. However, it has to be highlighted that in the conducted experiments the carbonator was operated with a synthetic flue gas without any SO_2. Due to SO_2 co-capture in the carbonator, the SO_2 carrying capacity of the limestone will be reduced affecting also the SO_2 capture in the calciner. This circumstance may be especially pronounced at low integration levels if a high sulphurous fuel is burned in the upstream process as commonly utilised in the cement industry, or a sulphurous rich flue gas is fed to the carbonator, respectively. However, based on the simulation of De Lena et al. [105][1] and the CEMCAP framework [17][2] a molar calcium to sulphur ratio of $473\,\mathrm{mol\,mol^{-1}}$ can be assumed for a calcium looping system with an integration level of 20 %. Hence, an even larger amount of calcium will be available for SO_2 capture in a full scale tail-end calcium looping cement plant integration. Arias et al. [20] investigated simultaneous CO_2 and SO_2 capture in an electrically heated pilot plant confirming that high SO_2 capture performance close to 100 % can easily be achieved in a calcium looping carbonator due to the surplus amount of CaO required for CO_2 capture. As a consequence, a caustic scrubber before the CPU may be forgone due to the efficient in-situ desulphurisation of the oxy-fuel calcium looping calciner. Furthermore, the carbonator can likely act as DeSOx unit of the cement plant.

---

[1]Fuel: Hard coal, $\gamma_S = 0.005\,\mathrm{kg\,kg^{-1}}$, $\mu_{CLK} = 0.737\,\mathrm{kg\,kg^{-1}}$, $H_i = 27\,\mathrm{MJ\,kg^{-1}}$
[2]$y_{SO_2,\text{flue gas}} = 90 \cdot 10^{-6}\,\mathrm{m^3\,m^{-3}}$

6.6.4   NO$_x$ emissions

The calcium looping calciner's NO$_x$ emissions were evaluated in a dedicated campaign and are compared with the work of Hofbauer [83] who investigated oxy-fuel combustion using a Colombian hard coal (El Cerrejon) at University of Stuttgart's fluidised bed pilot plant. The characteristics of the utilised coals differ only slightly in terms of sulphur content. The coal burnt within these calcium looping experiments (La Loma) had a sulphur content of 0.01 kg kg^{-1} compared to 0.005 kg kg^{-1} of the El Cerrejon coal. The nitrogen content of the two coals were 0.019 kg kg^{-1} (El Cerrejon) and 0.016 kg kg^{-1} (La Loma).

A significant increase in NO$_x$ emissions (figure 6.22) and NO$_x$ conversion (figure 6.23) was found when the experimental facility was operated in calcium looping mode (i.e. oxy-fuel combustion in lime bed) compared to oxy-fuel combustion with silica sand. The NO$_x$ emission ranged from $500 \cdot 10^{-6}$ m^3 m^{-3} to $750 \cdot 10^{-6}$ m^3 m^{-3} in case of calcium looping operation and from $55 \cdot 10^{-6}$ m^3 m^{-3} to $340 \cdot 10^{-6}$ m^3 m^{-3} in case of oxy-fuel combustion. Similarly, the NO$_x$ conversion increases. A somewhat stronger increase of NO$_x$ fuel conversion was observed compared to NO$_x$ emissions, due to the fact that the NO$_x$ concentration was diluted by CO$_2$ originating from sorbent calcination. While the oxy-fuel combustion experiments yielded fuel nitrogen conversions in the range of 0.01 mol mol^{-1} to 0.04 mol mol^{-1}, fuel nitrogen conversions between 0.10 mol mol^{-1} and 0.15 mol mol^{-1} were observed for the calcium looping experiments. A linear increase with excess oxygen is evident regardless of the operation mode. The linear increase with excess oxygen is well known and can be attributed to a decreased reducing zone in which NO$_x$ is reduced to N$_2$ on the surface of char particles [98, 175]. The enhanced formation of NO$_x$ at calcium looping operation conditions results from a combination of several effects: (i) the catalytic effect of CaO on NO$_x$ formation and N$_2$O decomposition [111, 175], (ii) the elevated temperature required for sorbent calcination [175] as well as (iii) an increased oxygen concentration intensifying the oxidation of nitrogen components while simultaneously decreasing the flue gas recirculation rate which eventually decreases the reducing zone of the calciner [83, 175]. The calcium looping experiments assessing NO$_x$ emissions were operated with an elevated temperature of approx. 920 °C and an inlet oxygen concentration of approx. 0.55 m^3 m^{-3}. Whereas the oxy-fuel experiments were conducted at combustion temperatures ranging from 837 °C to 916 °C and inlet oxygen concentrations ranging from 0.25 m^3 m^{-3} to 0.37 m^3 m^{-3}.

While the oxygen concentration can be adjusted by the flue gas recirculation rate, the other NO$_x$ promoting effects are inherent to the calcium looping operation. The high NO$_x$ emission of the calcium looping calciner will most certainly require a DeNOx unit before sequestration or utilisation. However, NO_x can easily be removed during the compression in the CPU [154, 179]. Elevated pressure favours the conversion from NO_x to NO_2. NO_2 has a high solubility allowing to form nitric acid by dissolving NO_2 in water after compression at approx. 30 bar in a dedicated contact column [154, 179]. Furthermore, long-term experiments at the Callide oxy-fuel plant in Australia regarding corrosion in the CPU unit of an oxy-fuel power plant reported no enhanced corrosion associated to NO_x or the forming nitric acid [115].

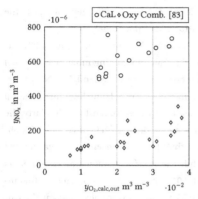

Figure 6.22: NO_x emissions (y_{NO_2}) vs. excess oxygen ($y_{O_2,calc,out}$) for calcium looping operation and oxy-fuel combustion with silica sand burning Colombian hard coal.

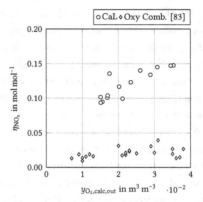

Figure 6.23: NO_x conversion (η_{NO_x}) vs. excess oxygen ($y_{O_2,calc,out}$) for calcium looping operation and oxy-fuel combustion with silica sand burning Colombian hard coal.

6.7 Conclusion on fluidised bed calcium looping CO_2 capture

Fluidised bed calcium looping investigating a wide range of operation conditions anticipated for CO_2 capture from cement plants has successfully been demonstrated at University of Stuttgart's fluidised bed pilot plant at the Institute of Combustion and Power Plant Technology. Overall, the calcium looping system operated as anticipated from the investigation of CO_2 capture from power plants. The increased sorbent make-up rates enabled by sorbent reutilisation in the cement clinker manufacturing process enhanced the sorbent's

CO_2 carrying capacity yielding carbonator CO_2 capture efficiencies of up to $0.98\,\mathrm{mol\,mol^{-1}}$ at carbonation temperatures of around 600 °C. For low integration levels between the cement plant and the calcium looping process (i.e. lower sorbent make-up rates and higher CO_2 load) the sorbent activity decreased and the CO_2 capture was limited by the amount of active CaO being fed to the carbonator. The experiments proved that the CO_2 capture efficiency can be adjusted using the circulation rate between the carbonator and the calciner or more specifically the active looping ratio. As the CO_2 concentration increases up to $0.30\,\mathrm{m^3\,m^{-3}}$ for low integration cases a significant reduction in fluidisation gas due to CO_2 absorption needs to be considered when designing the calcium looping carbonator to ensure proper fluidisation of the carbonation reactor. Besides, the carbonation temperature could easily be controlled by adjusting the cooling duty of the carbonator's bed heat exchanger.

High make-up rates of fresh sorbent imposed no issues on the oxy-fuel calciner. With minor exceptions regeneration efficiencies above $0.80\,\mathrm{mol\,mol^{-1}}$ were obtained throughout the experiments corresponding to a molar carbonate content of the exiting sorbent below $0.03\,\mathrm{mol\,mol^{-1}}$. Sorbent regeneration efficiencies showed no dependence on the calcination temperature, although only a narrow temperature range from 900 °C to 930 °C was investigated. Regardless of the operation conditions uniform temperature profiles without major hotspots could be attained by exploiting the good heat distribution characteristics of CFB reactors in combination with staged combustion and the temperature controlling effect of the endothermic calcination reaction. Heat release by combustion was controlled by staged oxidant feeding. The heat control mechanism also enabled operation with low flue gas recirculation rates or high inlet oxygen concentrations, respectively, without major deviation from the characteristic temperature profile. The calciner was successfully operated with inlet oxygen concentrations of up to $0.57\,\mathrm{m^3\,m^{-3}}$ corresponding to recirculation ratios of 20 %. Such mild and controlled calcination conditions favour a high sorbent activity as sintering is minimised and contribute therefore to the overall good performance of the calcium looping process.

Generally, the oxy-fuel operation of the CFB calciner is in agreement with the operation experience reported for oxy-fuel CFB combustors. CO emissions decreased with increasing excess oxygen. In-situ desulphurisation efficiencies close to 100 % were obtained in the oxy-fuel calciner due to the exceptional high amount of CaO being also available for desulphurisation. NO_x emissions increased linearly with increasing excess oxygen concentration and were significantly increased compared to oxy-fuel combustion due among other things to the catalytic effect of CaO on NO_x formation. The high NO_x concentration (up to $750 \cdot 10^{-6}\,\mathrm{m^3\,m^{-3}}$) will most certainly require a NO_x removal step during compression in the CPU. Based on the experiments a CO_2 purity around $0.95\,\mathrm{mol\,mol^{-1}}$ can be anticipated before the CPU of a commercial calcium looping system.

Design criteria originally developed for calcium looping CO_2 capture from power plants such as the active space time approach for the carbonator or the calciner proved to be also applicable for calcium looping CO_2 capture from cement plants. Based on the presented experiments the ground work for a fluidised bed calcium looping demonstration plant for CO_2 capture from cement plant has been designed within the CEMCAP project [41].

Chapter 7

Results - entrained flow calcium looping

In this chapter the experimental results regarding the characterisation of various sorbents for entrained flow calcium looping CO_2 capture application for cement plants are presented. Four European raw meal qualities as well as a high purity limestone fraction of one of the raw meals have been assessed regarding their calcium looping properties using two key parameters: the calcination degree (X_{calc}) and the recarbonation degree (X_{recarb}). The term recarbonation degree is used to describe the amount of CaO that binds CO_2 (i.e. that reacts to calcium carbonate) with respect to the available amount of CaO due to the fact that the entrained flow calcined sorbents were not fully calcined during the experiments.

$$X_{calc} = 1 - \frac{\Delta N_{CaCO_3,calc}}{N_{CaO,tot}} \tag{7.1}$$

$$X_{recarb} = \frac{\Delta N_{CaCO_3,carb}}{N_{CaO,available}} \tag{7.2}$$

7.1 Calcination and recarbonation degree

The calcination degrees obtained from the entrained flow characterisation are presented in figure 7.1, whereas the recarbonation degrees are presented in figure 7.2. Generally, the calcination degree increases with increasing residence time and increasing temperature (i.e driving force), whereas the recarbonation degree decreases with increasing residence time but is not affected by the calcination temperature.

Calcination

Some scattering of the calcination degree can be observed especially for the oxy-fuel experiments operated with a calcination temperature of 900 °C. The scattering can be ascribed to slight deviation of the calcination temperature. At calcination conditions close to the equilibrium slight changes of the calcination temperature significantly influence the calcination's driving force due to the exponential nature of the equilibrium CO_2 partial pressure. For instance the lower calcination degree of the 'Calcare' limestone fraction at 3.6 s (figure 7.1e) and the Rumelange raw meal at 3.3 s (figure 7.1d) originates from a slight temperature difference of approx. 8 K resulting in a reduced driving force of approx. 0.135 bar. The reduced driving force represents a decrease of 29 % for the Rumelange raw meal and of 37 % for the 'Calcare' limestone fraction as the reference calcination temperature was 905 °C and 910 °C, respectively. For the oxy-fuel experiments at 920 °C the effect is less pronounced as only minor deviations of the calcination temperature occurred between the experiments in the range of 2 K.

The Bilbao raw meal shows the best calcination performance of all investigated raw meals. Regardless of the calcination conditions or solid residence time a calcination degree of approx. 0.80 mol mol^{-1} was achieved. Increasing the temperature from 900 °C to 920 °C at oxy-fuel conditions led to a slight improvement of the calcination degree with increasing residence time (figure 7.1a). All the other raw meals as well as the limestone fraction of the Vernasca raw meal showed an enhancement of the calcination degree with increasing residence time and calcination temperature or driving force, respectively. The respective calcination degrees of the Geseke raw meal (figure 7.1c) and Rumelange raw meal (figure 7.1d) show a rise of the calcination degree associated with the elevated calcination temperature. For both calcination temperatures roughly the same upward gradient was observed. The limestone fraction of the Vernasca raw meal (figure 7.1e) yields constant calcination degrees for lower residence times. After a certain threshold the calcination degree increases with increasing residence time. The threshold after which the calcination degree rises shifts to lower residence times with increasing temperature. This might indicate that the heat influx into the particle was insufficient to provide the excessive temperature required to drive calcination. Similarly, the Vernasca raw meal (figure 7.2b) yielded a constant calcination degree of approx. 0.43 mol mol^{-1} for residence times below 3.5 s that increases to 0.56 mol mol^{-1} at 4.1 s for oxy-fuel calcination at 900 °C. However, for more calcining conditions (i.e. air calcination at 880 °C or oxy-fuel calcination at 920 °C) the calcination degree of the Vernasca raw meal increases steadily with increasing residence time. Presumably, the rather constant calcination degree of the Vernasca raw meal for lower residence times

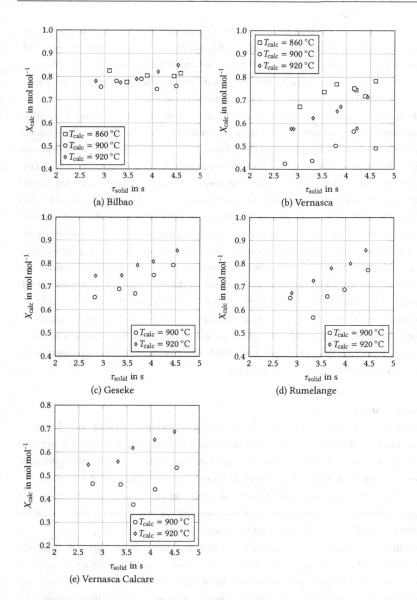

Figure 7.1: Calcination degree (X_{calc}) vs. residence time (τ_{solid}) for various raw meals and a raw meal limestone fraction calcined in entrained flow conditions. Oxy-fuel: 900 °C and 920 °C in 0.9 m^3 m^{-3} CO$_2$; air: 860 °C in 0.26 m^3 m^{-3} CO$_2$.

at 900 °C (oxy-fuel) as well as the overall lower calcination degree can be ascribed to the limestone fraction yielding lower calcination degrees due to its significantly higher $CaCO_3$ content. However, compared to the limestone fraction the Vernasca raw meal's calcination degree increases rather constantly indicating that the calcination performance is superimposed by the Vernasca raw meal's marl and limestone fractions.

Although the different sorbents show a quite diverse calcination performance it can be concluded that the calcination degree increases with increasing temperature and residence time. Nonetheless, the extent of the effectiveness of a change in temperature or residence time on the calcination degree or recarbonation degree differs substantially from sorbent to sorbent. On average, the Vernasca raw meal's calcination degree increased around $0.053\,mol\,mol^{-1}$ per second of solid residence time, whereas the Geseke raw meal showed an increase of $0.077\,mol\,mol^{-1}$ and the Rumleange an increment of $0.10\,mol\,mol^{-1}$ per second of residence time. At 920 °C in oxy-fuel conditions the Bilbao raw meal yielded a rise of $0.043\,mol\,mol^{-1}\,s^{-1}$ and the limestone fraction of the Vernasca raw meal achieved an increase of $0.085\,mol\,mol^{-1}\,s^{-1}$. An increase in residence time appears to be more important for raw meals containing a high purity limestone, as the time required to achieve full calcination of the limestone particles is prolonged for a given calcination rate. Contrary, the calcination degree of marl type raw meals increase stronger with increasing residence time, as such raw meals usually exhibit advantageous calcination properties. The increased driving force due to the elevated temperatures yielded similar improvements regardless of the raw meal origin or the sorbent composition.

Recarbonation

Overall, the recarbonation degree of the entrained flow calcined raw meals ranged between $0.5\,mol\,mol^{-1}$ and $0.3\,mol\,mol^{-1}$. Raising the calcination temperature from 900 °C to 920 °C at oxy-fuel conditions did not affect the recarbonation degree but with increasing solid residence time the recarbonation degree decreased generally. However, the Vernasca raw meal as well as its limestone fraction shows no significant dependency of the recarbonation degree on the residence time. Similarly, a constant recarbonation degree is observed for the Bilbao raw meal if calcined at oxy-fuel conditions.

The Bilbao raw meal (figure 7.2a) shows the highest recarbonation degree of all investigated raw meals. For the oxy-fuel experiments a rather constant recarbonation degree of $0.5\,mol\,mol^{-1}$ was found that decreased slightly at higher residence times for the 920 °C experiments. In case of air calcination the recarbonation degree decreased from $0.5\,mol\,mol^{-1}$ at 3.1 s to $0.43\,mol\,mol^{-1}$ at 4.4 s. The decrease of the recarbonation degree appears to be

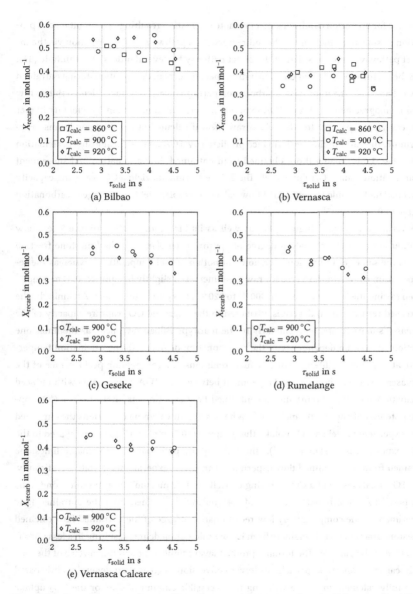

Figure 7.2: Recarbonation degree (X_{reacrb}) vs. residence time (τ_{solid}) of the entrained flow calcined sorbent samples. Oxy-fuel: 900 °C and 920 °C in $0.9\,\text{m}^3\,\text{m}^{-3}$ CO_2; air: 860 °C in $0.26\,\text{m}^3\,\text{m}^{-3}$ CO_2; carbonated at 650 °C in $0.10\,\text{m}^3\,\text{m}^{-3}$ CO_2 and $0.10\,\text{m}^3\,\text{m}^{-3}$ H_2O

favoured for operation conditions that promote calcination resulting in a simultaneous promotion of silicate formation via the indirect reaction pathway. Belite formation via the indirect pathway overcomes that of the direct pathway by several orders of magnitude [11], thus, being the most relevant reaction with respect to sorbent deactivation during entrained flow calcination. The improved recarbonation performance of oxy-fuel calcined Bilbao raw meal is in agreement with the TGA experiments presented in section 5.2 yielding higher CO_2 carrying capacities for the Bilbao raw meal if calcined at oxy-fuel conditions. Furthermore, the carbonation performance predicted by TGA analysis using short calcination times (i.e. 1 min) match those obtained from entrained flow calcination and subsequent recarbonation using TGA fairly well. The TGA experiments yield a CO_2 carrying capacity of $0.4\,mol\,mol^{-1}$ while the entrained flow calcined samples yield an average recarbonation degree of $0.5\,mol\,mol^{-1}$.

The Vernasca raw meal (figure 7.2b) as well as its limestone fraction (figure 7.2e) show a rather constant recarbonation degree of approx. $0.4\,mol\,mol^{-1}$. The limestone fraction shows the same recarbonation behaviour for both oxy-fuel temperatures, whereas the recarbonation degree of the Vernasca raw meal increased slightly by approx. $0.04\,mol\,mol^{-1}$ when raising the temperature from 900 °C to 920 °C at oxy-fuel conditions. Assumingly, the increased recarbonation degree is attributed to the improved CO_2 capture capacity of the Vernasca's marl component, similarly to the marltype Bilbao raw meal, as the limestone fraction showed no dependency of the recarbonation degree with the calcination temperature at oxy-fuel conditions. For oxy-fuel conditions, the carbonation performance of the Vernasca raw meal shows a good agreement between the TGA experiments with reduced calcination time (i.e. 1 min) and the entrained flow experiments. Both yield a CO_2 capture potential of approx. $0.4\,mol\,mol^{-1}$. Whereas, for air calcining conditions the entrained flow experiments yield a CO_2 uptake that is approx. $0.26\,mol\,mol^{-1}$ lower compared to the TGA experiments ($0.64\,mol\,mol^{-1}$). The CO_2 capture potential of the Vernasca limestone obtained from the entrained flow experiments and TGA experiments deviate significantly. The TGA analyses yield a CO_2 carrying capacity of $0.65\,mol\,mol^{-1}$ for oxy-fuel conditions compared to a recarbonation degree of $0.41\,mol\,mol^{-1}$ obtained from the entrained flow experiments. The comparatively low recarbonation degree of the entrained flow calcined limestone fraction might originate from its low calcination degree resulting in a less porous structure of the particle. The forming product layer might seal the outer surface of the partially calcined limestone particles at lower conversion degrees than generally anticipated for a fully calcined limestone reducing the accessible amount of CaO or the CO_2 uptake potential, respectively.

The Geseke raw meal (figure 7.2c) and the Rumelange raw meal (figure 7.2d) both yield similar recarbonation degrees for the two investigated calcination temperatures at oxy-fuel conditions. Their recarbonation degree decreases from $0.45\,\mathrm{mol\,mol^{-1}}$ and $0.44\,\mathrm{mol\,mol^{-1}}$ at 2.8 s to $0.33\,\mathrm{mol\,mol^{-1}}$ and $0.32\,\mathrm{mol\,mol^{-1}}$ at 4.5 s, respectively. The obtained recarbonation degrees of these two raw meals are also in adequate agreement with the CO_2 carrying capacities obtained from TGA experiments with 1 min calcination.

The slight decrease of the CO_2 capture performance with increasing calcination time of the raw meal based sorbents originates most likely from progressing belite conversion as sintering is less pronounced for temperatures below 950 °C and belite formation starts to occur at temperatures around 700 °C [76, 114]. The rather constant recarbonation degree of the high purity limestone further supports this hypothesis as belite formation is negligible due to the limestone's low silicon content.

Overall, the recarbonation degrees obtained from entrained flow calcined raw meals are in good agreement with the CO_2 carrying capacity obtained from the cyclic TGA experiments operated with a short calcination time of 1 min. However, when comparing these two parameters it has to be highlighted that the entrained flow calcined samples are not fully calcined, whereas the TGA experiments hold excess time after full calcination is achieved allowing longer belite formation. The additional reaction time required to achieve full calcination will allow belite formation to progress leading to slightly lower recarbonation degrees of the fully entrained flow calcined raw meal based sorbent.

7.2 Influence of fuel combustion on the calcination and recarbonation performance

The effect of natural gas combustion inside the calciner on key calcium looping parameters (X_{calc} und X_{recarb}) has been assessed using two different sets of experiments. Initially, the effect of fuel combustion has been assessed at oxy-fuel conditions with a calcination temperature of 900 °C. To reduce local hotspots and to avoid emergency shutdowns due to local temperature peaks above 1050 °C, staged fuel feeding was applied at air calcination conditions (880 °C). The natural gas was evenly distributed between the gas distributor at the reactor bottom (0 m) and a fuel injection lance installed at 2.8 m. Figure 7.3 presents the effect of fuel combustion on the calcination and recarbonation degree. The corresponding temperature profiles are depicted in figure 7.4. From figure 7.3 a strong improvement of the calcination degree with increasing fuel feeding rate is evident regardless of the reaction atmosphere. Moreover, an increase of the calcination degree is obvious when changing from oxy-fuel to air calcination (i.e. reducing the CO_2 partial pressure). For air calcination, the calcination degree increased from $0.52 \, mol \, mol^{-1}$ without fuel combustion to $0.74 \, mol \, mol^{-1}$ at $0.2 \, m^3 \, h^{-1}$ and further rose to $0.90 \, mol \, mol^{-1}$ at a fuel feeding rate of $0.6 \, m^3 \, h^{-1}$. Concurrently, the recarbonation degree decreased with increasing calcination degree or fuel feeding rate, respectively. Equal recarbonation degrees were observed for air and oxy-fuel calcination regardless of the fuel feeding rate. Since the Vernasca raw meal's recarbonation degree proved to be constant for all calcination degrees the observed sorbent deactivation can likely be attributed to fuel combustion. The enhanced fuel feeding led to a local hotspot at the bottom of the reactor where the natural gas was injected. Without fuel staging hotspots above 1000 °C occurred, whereas with staged fuel feeding hotspots up to 980 °C shortly above the feeding stage (i.e. 0 m and 2.8 m) were formed. Besides, due to the intensified combustion the heat influx into the raw meal particles improves. Thus, the calcination degree increases with increasing fuel feeding rate. The enhanced deactivation of the raw meal's CO_2 uptake potential is likely to originate from a combination of intensified sintering as well as enhanced belite formation. The share of sintering and belite formation on the observed sorbent deactivation is hard to distinguish since both effects are promoted by increasing temperature. However, since sintering is known to be of subordinate importance at temperatures below 950 °C [76], it is likely that the enhanced deactivation originated primarily form belite formation as more CaO is available to form belite. Alonso et al. [11] reported that belite formation via the indirect pathway (i.e. CaO) exceeds the direct pathway at calcining conditions by several orders of magnitude. Consequently, the rise in the calcination degree indirectly promotes belite formation. Moreover, the belite reaction rate increases substantially with increasing temperature [11]. The local hotspots after fuel injection facilitate belite formation as local melting is more likely to occur in the hotspot zones. Therefore, it can be assumed that the observed sorbent deactivation with increasing fuel feeding originates from enhanced belite formation via the indirect pathway promoted by the formation of local hotspots. Hence, heat control measures such as oxidant or fuel staging should be employed to monitor and control the heat release and to avoid local hotspots and minimise silicate formation.

The strong increase of the calcination degree associated with fuel combustion implies that heat transfer solely by radiation of the reactor walls was insufficient to achieve adequate calcination within the residence time of typical entrained flow calciner (3 s to 6 s). Heat transfer by radiation was limited due to the small temperature differences of approx. 100 K between the reactor wall and the particle. However, the LEILAC technology (i.e. indirect cal-

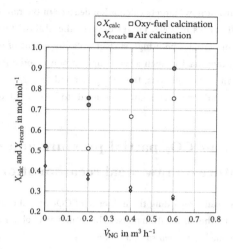

Figure 7.3: Calcination degree (X_{calc}) and recarbonation degree (X_{recarb}) vs. natural gas feeding rate ($\dot{V}_{NG}$) for the Vernasca raw meal at air conditions ($\tau_{solid} \approx 4.1\,s$) and oxy-fuel conditions ($\tau_{solid} \approx 3.9\,s$).

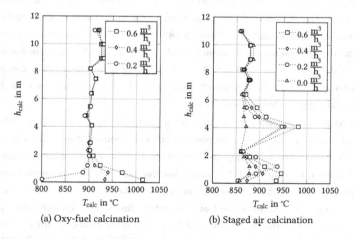

(a) Oxy-fuel calcination   (b) Staged air calcination

Figure 7.4: Temperature profile for the investigation of the influence of fuel feeding on the calcination and recarbonation degree of the Vernasca raw meal.

cination in countercurrent flow) targets calcination degrees in the range of $0.95 \, mol \, mol^{-1}$ with temperature differences of $100 \, K$ [173]. Besides, the LEILAC technology is known to produce magnesium particles with high surface areas by means of indirect calcination [148, 149]. Calcining raw meal in such a way might produce sorbent with an enhanced carbonation performance. Nonetheless, the CO_2 generated to supply heat to the indirect calciner represents an additional CO_2 source that needs to be accounted for.

7.3   Influence of CO_2 partial pressure and carbonation temperature on raw meal recarbonation

This section presents results regarding the influence of CO_2 partial pressure and carbonation temperature during the carbonation phase. The results were obtained by TGA analysis of entrained flow calcined raw meal samples using the analysis routine described in subsection 4.3.3. The Vernasca raw meal was chosen as a representative of a heterogeneously mixed raw meal, whereas the Bilbao raw meal was selected as a representative of marl type raw meals. The entrained flow calcined samples calcined at air conditions (880 °C) as well as oxy-fuel conditions (900 °C, 920 °C) have been evaluated in the TGA at three carbonation temperatures (600 °C, 650 °C and 700 °C) using three different CO_2 partial pressures or CO_2 concentration ($0.05 \, m^3 \, m^{-3}$, $0.10 \, m^3 \, m^{-3}$ and $0.20 \, m^3 \, m^{-3}$). The respective results of the Bilbao raw meal are presented in figure 7.5 on the left side while the results referring to the Vernasca raw meal are presented on the right side of figure 7.5. Due to the low driving force of the $0.05 \, m^3 \, m^{-3}$ experiments at a carbonation temperature of 700 °C carbonation did not complete within the 15 min carbonation phase for the Vernasca raw meal. Hence, these results are presented as grey triangles.

No significant impact of the CO_2 partial pressure on the CO_2 uptake was observed for both raw meals in the investigated partial pressure range (i.e. below 0.2 bar). An improvement of the CO_2 uptake with increasing carbonation temperature is evident for all investigated cases. However, if calcined at 920 °C in oxy-fuel conditions no further improvement was observed when raising the carbonation temperature from 650 °C to 700 °C (figure 7.5e and 7.5f). Similarly, the enhancing effect of temperature diminishes for carbonation temperatures above 650 °C if the Bilbao raw meal is calcined at air conditions (figure 7.5a). For oxy-fuel conditions and a calcination temperature of 900 °C, the Bilbao raw meal shows a linear increase from $0.45 \, mol \, mol^{-1}$ at 600 °C to $0.67 \, mol \, mol^{-1}$ at 700 °C. Similarly, the Vernasca raw meal yields a linear increase from approx. $0.25 \, mol \, mol^{-1}$ at

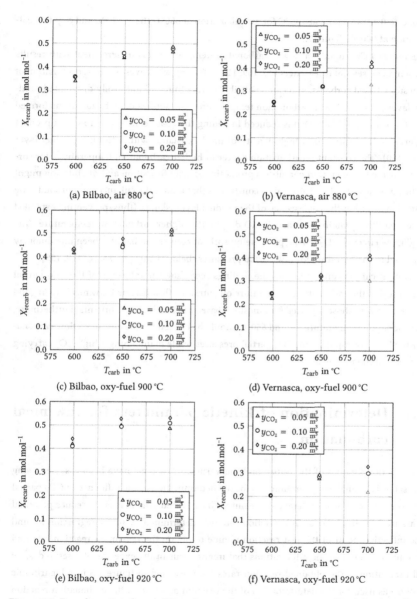

Figure 7.5: Recarbonation degree (X_recarb) vs. carbonation temperature (T_carb) for the Bilbao raw meal (left) and Vernasca raw meal (right) calcined at various operation conditions.

600 °C to 0.40 mol mol^{-1} at 700 °C for air calcination at 880 °C (figure 7.5b) and oxy-fuel calcination at 900 °C (figure 7.5d).

The reduced CO_2 uptake at lower carbonation temperatures is in agreement with results presented by several researchers [45, 109, 110, 120, 133], who investigated recarbonation of limestone based sorbents, and can be attributed to a thin and uniformly distributed product layer [110]. At lower carbonation temperatures, the solid state diffusion of the formed $CaCO_3$ or its ions (Ca^{2+}, O^{2-}) is reduced impeding the coalescence and growth of product layer islands resulting in a high density of product islands with a thinner product layer [110, 120]. The constant recarbonation degree at higher carbonation temperatures for sorbent calcined at operation conditions promoting silicate formation and/or sintering might indicate some kind of morphological constraints that inhibit the enhancement of solid state diffusion or rather the coalescence of the product layer islands. Thus, preventing extended access to the remaining unreacted CaO despite the higher carbonation temperature. The negligible effect of CO_2 partial pressure on sorbent conversion has also been unanimously reported in the literature for low partial pressures (below 1 bar) [29, 74, 76, 133, 155].

Overall, it can be concluded that the findings regarding the influence of the CO_2 partial pressure and the carbonation temperature on entrained flow calcined raw meal are in good agreement with those reported for limestone based calcium looping sorbents. Furthermore, a carbonation temperature around 650 °C should be chosen for the entrained flow carbonator to allow low equilibrium CO_2 partial pressures while maintaining a high CO_2 carrying capacity of the raw meal.

7.4 Determination of kinetic parameters for raw meal carbonation

The reaction kinetics of the Bilbao and the Vernasca raw meal have been assessed using the same TGA analysis programme as for the determination of the influence of CO_2 partial pressure and carbonation temperature but with a reduced sample mass to reduce potential mass transfer limitations. The reaction rate was determined by linear regression around the inflection point of the fast reaction regime to obtain the maximum reaction rate and to minimise any transport limitations that might occur in the experimental setup despite all precautions. The obtained reaction rates are presented in figure 7.6 in a logarithmic pattern as used for the determination of the reaction order (n) of the carbonation reaction with respect to CO_2 using linear regression (eq. 7.3). For the logarithmic calculation the

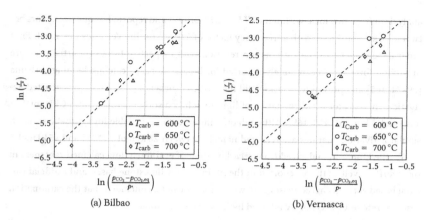

(a) Bilbao   (b) Vernasca

Figure 7.6: Carbonation reaction rate (r) against CO_2 partial pressure ($p_{CO_2} - p_{CO_2,eq}$) for the Bilbao raw meal (a) and the Vernasca raw meal (b). The fit determined by linear regression is depicted as dashed lines.

reaction rate was normalised using a normalising reaction rate (r^*) of $1\,mol\,mol^{-1}\,s^{-1}$ and a normalising pressure of $1.013\,bar$.

$$\ln(r) = n \cdot \ln(p_{CO_2} - p_{CO_2,eq}) + \ln(k_{carb} \cdot X_{CO_2,max}) \qquad (7.3)$$

The linear increase of the logarithmic carbonation reaction rate with the logarithmic driving force is evident for both the Bilbao and the Vernasca raw meal. Their respective reaction order was determined by linear regression of the logarithmic driving force ($p_{CO_2} - p_{CO_2,eq}$) and the logarithmic reaction rate (r). A reaction order of 0.99 with a normalised root-mean-square deviation (NRMSD) of 0.075 was determined for the Bilbao raw meal, whereas a reaction order of 0.92 with a NRMSD 0.093 was obtained for the Vernasca raw meal. The frequency factor (k_0) and activation energy (E_A) was determined assuming an Arrhenius type dependency of the carbonation constant. The activation energy was derived by linear regression against the inverse temperature using the logarithmic form of the Arrhenius equation. The Bilbao raw meal yielded an activation energy of $11.4\,kJ\,mol^{-1}$ and a frequency factor of $1.8\,s^{-1}$, whereas the Vernasca raw meal yielded an activation energy of $11.5\,kJ\,mol^{-1}$ and a frequency factor $2.1\,s^{-1}$. These obtained activation energies are slightly smaller than those proposed by Fernandez et al. [65] for the carbonation of raw meal based sorbents ($15\,kJ\,mol^{-1}$). Accordingly, the determined frequency factor differs slightly. Fernandez et al. [65] investigated among other raw meals the Vernasca raw meal using TGA analysis of fresh raw meal within the *CLEANKER* project and reported frequency factors between $4.8\,s^{-1}$ and $7.6\,s^{-1}$ with a frequency factor of $5.06\,s^{-1}$ for the Vernasca raw meal. First order dependencies of the carbonation reaction with respect to the CO_2 partial pressure or driving force have been widely reported for limestone based calcium looping sorbents [29, 74, 165] and proved to be also valid for artificial sorbents [186, 187]. The reported activation energies for the carbonation of limestone based sorbents range from $0\,kJ\,mol^{-1}$ to $32\,kJ\,mol^{-1}$ [29, 151, 152, 165]. Although only minor data is available for synthetic sorbent, similar activation energies are reported namely between $28\,kJ\,mol^{-1}$ [187] and $45\,kJ\,mol^{-1}$ [186]. The kinetic data derived from recarbonation of entrained flow calcined raw meal in this work as well as those reported in the literature for limestone based and artificial calcium based sorbents agree reasonably well. Thus, it can be concluded that the carbonation reaction rate of CaO is hardly affected by its sorbent origin.

7.5   Conclusion on entrained flow calcium looping CO_2 capture

A wide range of raw meal based sorbents have been assessed for the application in an entrained flow calcium looping system. Generally, calcination performance and recarbonation degree behaved as anticipated. Calcination improved with increasing residence time and calcination temperature. Furthermore, heat release by fuel combustion inside the calciner resulted in an increased calcination degree with the draw back of enhanced sorbent deactivation due to local hotspots. As a consequence, heat control measures such as staged fuel or oxidant feeding should be employed to maintain a more active sorbent. Based on the results, an increased temperature slightly above $920\,^\circ C$ in case of oxy-fuel calcination is required to achieve a calcination performance comparable to those of state of the art precalciners operated with air combustion. The recarbonation degree decreased slightly with increasing residence time. Higher calcination temperatures did not affect the raw meal's CO_2 capture performance. Belite formation that inevitably occurs during calcination will bind CaO and therefore reduce the sorbent's CO_2 capture potential. With sufficiently fast calcination times the extend of sorbent deactivation can be limited.

Similar to the recarbonation of limestone based sorbents, the raw meal's CO_2 carrying capacity increased with increasing carbonation temperature. For optimal CO_2 capture a carbonation temperature around $650\,^\circ C$ is suggested. The CO_2 partial pressure did not affect the sorbent's CO_2 uptake but the carbonation reaction rate. A first order dependency of the carbonation reaction with respect to the CO_2 partial pressure difference (i.e. driving force) was found for the two selected raw meals. Furthermore, TGA experiments appear to be able to adequately describe the CO_2 carrying capacity of entrained flow calcined raw meals if sufficiently fast heat up rates and short calcination times are employed. Overall, it can be concluded that raw meal based sorbents perform analogous to limestone based sorbents, but are superimposed by belite formation that cause an additional reduction of the CO_2 capture capacity. Thus, an optimised calcination reaction time is required to minimise sorbent deactivation by belite formation.

Chapter 8

Summary and conclusions

There is an imminent need to reduce CO_2 emission from cement clinker manufacturing to mitigate anthropogenic climate change and limit the average global temperature rise to below 2 °C. Since alternative, less carbon intensive production routes, such as raw meal calcination by electrolysis or heat supply by plasma torches and microwave heating, are far from market maturity the IPCC and IEA consider that the application of CCS technologies in the cement industry is inevitable especially when considering the increasing demand for cement of the growing emerging economies. A promising CCS candidate, the calcium looping CO_2 capture process, has been comprehensively assessed in this thesis. This CO_2 capture technology appears to be especially suitable for the cement industry as both processes rely on calcium containing solids enabling new beneficial operation conditions for the calcium looping process. Within the framework of this thesis various possibilities to integrate calcium looping CO_2 capture into the cement clinker manufacturing process have been developed in collaboration with research partners and experimentally assessed using the infrastructure at the University of Stuttgart.

Generally, fluidised bed reactors or entrained flow reactors are feasible to be employed in case of cement plant application. For calcium looping, fluidised bed reactors represent the more mature reactor concept that has proven its feasibility for power plant application. The employment of entrained flow reactors is a novelty that enables a more direct integration of the calcium looping CO_2 capture step into the cement manufacturing process. To characterise the integration extent between the two processes the term integration level is introduced to define the calcium looping boundary conditions such as sorbent make-up and CO_2 load based on the cement plant's clinker production. Fluidised bed integration options are generally more energy demanding since the purge is expected to be cooled and ground to a finer particle size suitable for clinker burning in the cement kiln, whereas purge from entrained flow application can be directly used as feedstock for clinker production. Energy recuperation from purge cooling will increase the electricity generation of the calcium looping steam cycle.

A broad range of potential calcium looping sorbents have been assessed regarding their cyclic CO_2 carrying capacity using thermogravimetric analysis. From this assessment it can be concluded that silica containing sorbents (i.e. marlstoneish sorbents) suffer a severe initial degradation in their CO_2 carrying capacity that can be ascribed to belite formation. The degradation increases with increasing silica content but also depends on the homogeneity of the sorbent's calcium silicon aggregation.

The presence of steam enhances the CO_2 carrying capacity of both limestone and marlstone type sorbents, whereas high CO_2 partial pressures promote sintering of limestones while belite formation is hindered. Reduced calcination times lessen the extent of sorbent deactivation. Belite formation and sintering have less time to progress since the calcination reaction surpasses both deactivation reactions. For oxy-fuel conditions with longer calcination times (i.e. 10 min), the cyclic CO_2 carrying capacity of all investigated raw meals can be adequately described using the deactivation model of Grasa et al. with a deactivation constant of 3.29 and a residual sorbent activity of $0.10 \, mol \, mol^{-1}$ or using the model of Ortiz et al. with a CO_2 carrying capacity in the first cycle of $0.30 \, mol \, mol^{-1}$, a deactivation constant of 0.44 and a residual activity of $0.076 \, mol \, mol^{-1}$. For reduced calcination times of 1 min the raw meals' CO_2 carrying capacity can be described with the model of Grasa et al. using a deactivation constant of 2.27 and a residual sorbent activity of $0.15 \, mol \, mol^{-1}$ or with the model of Ortiz et al. using an initial sorbent activity of $0.39 \, mol \, mol^{-1}$, a deactivation constant of 0.28 and residual CO_2 carrying capacity of $0.09 \, mol \, mol^{-1}$.

The investigations concerning entrained flow calcium looping showed that the raw meal's CO_2 capture performance was hardly affected by the calcination temperature (in the investigated temperature range) but local hotspots caused significant sorbent deactivation. Hence, monitoring and controlling the heat release in the oxy-fuel calciner is crucial to maintain a high sorbent CO_2 carrying capacity. Furthermore, the recarbonation performance of entrained flow calcined raw meal is in agreement with recarbonation behaviour of limestone based sorbents known from power plant related research. A first order dependency of the carbonation reaction regarding the CO_2 partial pressure difference (i.e. driving force) was found for two raw meals representing a marl type raw meal and heterogeneously mixed raw meal. Also, carbonation temperatures around 650 °C proved beneficial due to the sorbent's enhanced CO_2 carrying capacity and the low achievable CO_2 partial pressure in the carbonation reactor.

Within this thesis, University of Stuttgart's 200 kW pilot plant has been used to successfully demonstrate fluidised bed calcium looping CO_2 capture at operation conditions representing those anticipated for high and low integration levels between a cement plant and the calcium looping process comprising make-up ratios up to 0.9 mol mol^{-1} and flue gas CO_2 concentrations up to 0.30 m^3 m^{-3}. For high integration levels, CO_2 capture in the carbonator was limited by the achievable equilibrium CO_2 concentration yielding CO_2 capture efficiencies of up to 98 % due to the significantly increased sorbent activity. At lower integration levels CO_2 capture was limited by the amount of active CaO being fed to the carbonator. These experiments proved that the CO_2 capture efficiency can be adjusted using the circulation rate between the carbonator and the calciner or more specifically the active looping ratio.

Smooth calciner operation over the wide range of operation conditions such as sorbent make-up rates, calcination temperature or flue gas recirculation rate has further been achieved. The beneficial heat control properties of the CFB calciner in combination with staged fuel combustion and the temperature controlling effect of endothermic calcination reaction allowed operation with oxygen inlet concentrations up to 0.57 m^3 m^{-3} without major deviation from the calciner's characteristic temperature profile. The sorbent regeneration efficiency could be adequately described by the active space time approach of the calcination reactor. Generally, molar carbonate contents below 0.03 mol mol^{-1} were achieved corresponding to regeneration efficiencies between 0.80 mol mol^{-1} and 0.98 mol mol^{-1} depending on the sorbent make-up rate. Based on the experiments, a CO_2 purity of the flue gas exiting the calciner of 0.95 m^3 m^{-3} can be anticipated for commercial calcium looping units before the CPU. The CO_2 rich flue gas is primarily diluted with excess oxygen. Sulphurous components were completely absent due to in-situ desulphurisation. NO_x formation in the calcium looping calciner is significantly increased due to the presence of CaO catalysing the oxidation of nitrogenous components. NO_x concentration increased with excess oxygen and ranged between $500 \cdot 10^{-6}$ m^3 m^{-3} and $750 \cdot 10^{-6}$ m^3 m^{-3}. The high NO_x concentration will most certainly require a NO_x removal unit during CO_2 compression in the CPU.

Overall, it can be concluded that raw meal based sorbents behave similar to limestone based sorbents but are superimposed by an additional deactivation mechanism, namely belite formation. Sorbent deactivation can be minimised by reducing the calcination time and by avoiding local hotspots. Both, the active space time approach for the carbonator and the calciner can be used to design a fluidised bed calcium looping CO_2 capture system for cement plant application. The beneficial operation conditions associated with calcium looping purge reutilisation in the cement plant facilitate high CO_2 capture efficiencies making the calcium looping technology a viable option for efficient CO_2 capture from cement plants.

Recommendation for further work:

A major draw back of the fluidised bed calcium looping options is the need to cool down the purge and grind it to the fineness required for maintaining a certain clinker quality. However, fluidised bed operation with an average particle diameter in the range of $150\,\mu m$ is feasible. Hence, it should be verified whether a grounding step is necessary to maintain the clinker quality or if the direct reutilisation of such 'small sized' purge would be feasible. Especially for lower integration levels a direct use might be possible as in that case the calcium looping purge represents a lower share of the final cement clinker.

Aside from that, a hybrid oxy-fuel calciner, calcining a coarser raw meal fraction utilised as sorbent within the fluidised bed calcium looping system as well as calcining a fine raw meal fraction that is directly fed to clinker manufacturing, could significantly reduce the CO_2 load towards a tail-end fluidised bed calcium looping system as all CO_2 originating from raw meal calcination would be captured by oxy-fuel calcination. Despite the lower effective make-up rates (i.e. coarse particles) high make-up ratios could be realised in such a way. The operability of such a system should be assessed as it offers easy retrofitability with an efficient energy integration. Regardless, the fluidised bed calcium looping technology is ready to be implemented as a demonstration unit capturing CO_2 from a real cement plant to gather long-term operation experience before scaling up to a full size, first of a kind calcium looping plant. The ground work for designing and sizing such a fluidised bed calcium looping demonstration plant has already been performed within the *CEMCAP* project.

The less energy demanding entrained flow options still need to be demonstrated in full loop operation instead of assessing the calcination and carbonation separately as conducted in this work. The novel entrained flow calcium looping technology further holds uncertainties especially related to the carbonation performance under realistic entrained flow conditions. However, these issues are currently being addressed within the *CLEANKER* project using an entrained flow demonstration calcium looping plant treating around $2\,\%$ of a cement plant's flue gas.

Appendix A

Solid materials

A.1 Coal utilised within the fluidised bed experiments

Two Columbian hard coals were used as fuel within the fluidised bed experiments. Since the El Cerrejon coal was not dispatchable after the initial experimental campaign, the La Loma coal, located around 250 km south west of El Cerrejon mine, was employed. The coals were dried to a water content of approx. $0.07 \, \text{kg} \, \text{kg}^{-1}$ using warm air lances. Subsequently, the coals were crushed using a 4 mm sieve. Their chemical composition is presented in section 4.2 table 4.4, whereas their particle size distribution is depicted in figure A.1.

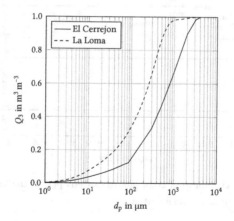

Figure A.1: Cumulative particle size distribution of the coal utilised within the experimental assessment of calcium looping CO_2 capture using fluidised bed reactors.

A.2   Utilised sorbents - limestones and marlstones

Additional information about the various sorbents assessed within this thesis are comprised below. The utilised sorbents are filed according to the classification of Correns [44] in figure A.2. For the classification only the calcium, silicon and alumina content are contemplated. Other components were considered to be inert for calcium looping operation. The respective content has been corrected by the amount of inerts. Furthermore, the calcium and silicon content were corrected by the share of Al_2O_3 to yield a reference composition comprising solely of $CaCO_3$ and SiO_2. The chemical composition of the various sorbents can be found in section 4.1 , table 4.1 (limestones) and table 4.2 (raw meals and raw meal components). Subsequently, the particle size distributions of the various materials are presented.

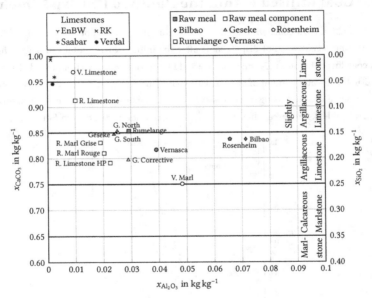

Figure A.2: Sorbent classification according to Correns [44] of the various utilised raw meals, raw meal components and limestones.

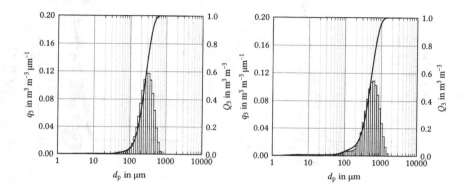

Figure A.3: Particle size distribution of the Lhoist limestone from Messinghausen (RK). Size cut: 100 μm to 300 μm.

Figure A.4: Particle size distribution of the Lhoist limestone from Messinghausen (RK). Size cut: 300 μm to 700 μm.

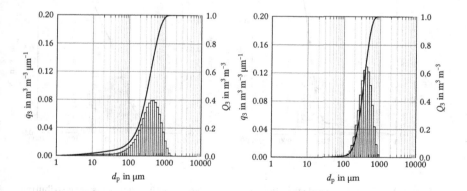

Figure A.5: Particle size distribution of the Verdal limestone.

Figure A.6: Particle size distribution of the EnBW limestone.

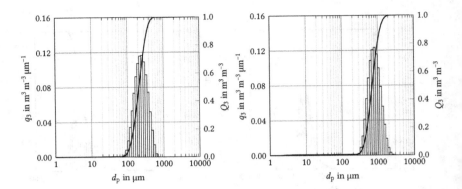

Figure A.7: Particle size distribution of the Saabar limestone.

Figure A.8: Particle size distribution of the limestone fraction of the Vernasca raw meal, crushed and sieved. Size cut: 420 μm to 1000 μm.

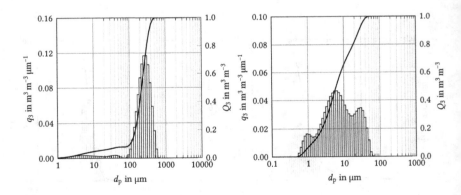

Figure A.9: Particle size distribution of the limestone fraction of the Vernasca raw meal, crushed and sieved. Size cut: 100 μm to 420 μm.

Figure A.10: Particle size distribution of the limestone fraction of the Vernasca raw meal, crushed and sieved below 100 μm.

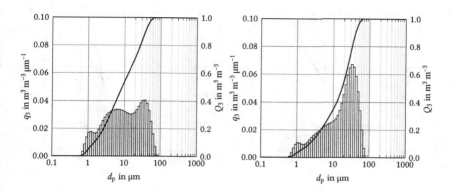

Figure A.11: Particle size distribution of the Vernasca raw meal.

Figure A.12: Particle size distribution of the marl fraction of the Vernasca raw meal, crushed and sieved below 100 μm.

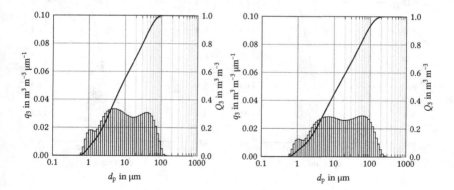

Figure A.13: Particle size distribution of the Bilbao raw meal.

Figure A.14: Particle size distribution of the Rosenheim raw meal.

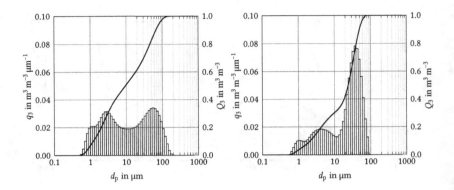

Figure A.15: Particle size distribution of the Geseke raw meal.

Figure A.16: Particle size distribution of the Geseke raw meal fraction 'North', crushed and sieved below 100 µm.

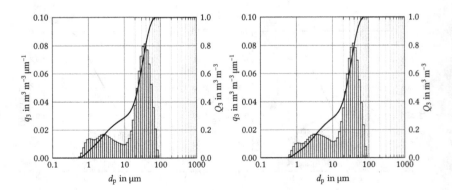

Figure A.17: Particle size distribution of the Geseke raw meal fraction 'South', crushed and sieved below 100 µm.

Figure A.18: Particle size distribution of the 'Corrective' fraction of Geseke raw meal, crushed and sieved below 100 µm.

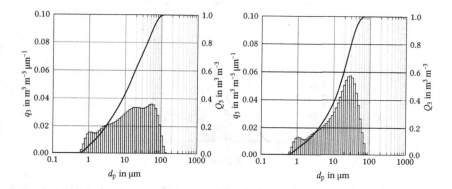

Figure A.19: Particle size distribution of the Rumelange raw meal.

Figure A.20: Particle size distribution of the limestone fraction of the Rumelange raw meal, crushed and sieved below 100 µm.

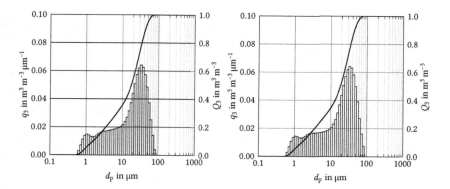

Figure A.21: Particle size distribution of the limestone fraction 'HP' of the Rumelange raw meal, crushed and sieved below 100 µm.

Figure A.22: Particle size distribution of the marl fraction 'Rouge' of the Rumelange raw meal, crushed and sieved below 100 µm.

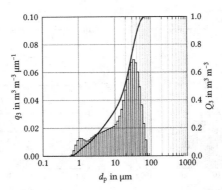

Figure A.23: Particle size distribution of the marl fraction 'Grise' of the Rumelange raw meal, crushed and sieved below 100 μm.

Appendix B

Calcium looping cement plant integration

Depending on the cement plant's boundary conditions various calcium looping integration options are feasible. In theory, the integration options can be distinguished by the number of calcination steps (i.e. oxy-fuel calcination and/or air calcination), the type of reactor employed for the calcium looping system (i.e. fluidised bed (FB) or entrained flow (EF) reactors), the type of sorbent used within the calcium looping process (i.e. utilised raw meal components or raw meal mix) and furthermore, the amount of sorbent fed to the calcium looping system with respect to the clinker production (i.e. integration level). The type of solid material best utilised as calcium looping sorbent depends on the available stone qualities. Preferably, the components with the best capture performance are employed. Accordingly, limestones are favoured over marlstones in heterogeneous meals as they generally exhibit a higher CO_2 carrying capacity. Obviously, sorbent selection is only feasible for integration levels below 100 %.

B.1 Back-end calcium looping option using entrained flow reactors

Like all tail-end solutions, the back-end entrained flow integration is characterised by a double calcination. A fraction of the raw meal, that is used as sorbent for the calcium looping CO_2 capture process, is calcined at oxy-fuel conditions. Therefore, this share of raw meal CO_2 is immediately captured. The remaining share of raw meal is pre-heated in the cement plant's suspension pre-heater cascade and calcined at air condition. The

CO_2 released during air calcination as well as the CO_2 generated by fuel combustion is fed to the carbonator to be captured. The share of raw meal used as sorbent within the calcium looping system can be independently chosen. With increasing sorbent share the calcination duty is shifted towards the calcium looping calciner and the CO_2 load towards the carbonator decreases. For high integration levels, the cement plant's pre-calciner might be omitted if the kiln flue gas' sensible heat is sufficient to preheat and fully calcine the partially calcined raw meal fed to the suspension pre-heater. Since entrained flow reactors are used the sorbent is already milled to the required size to produce clinker. Thus, it can be fed to the cement plant without further processing.

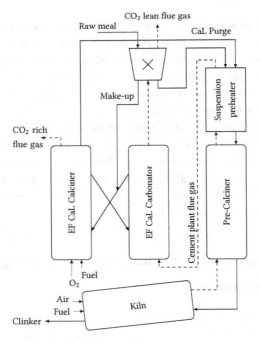

Figure B.1: Schematic of a tail-end calcium looping CO_2 capture option using entrained flow reactors (EF).

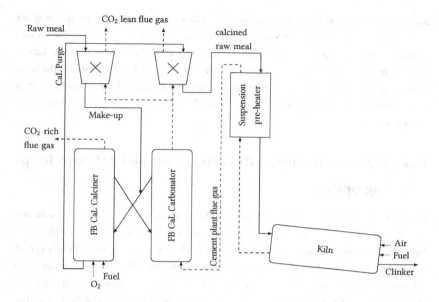

Figure B.2: Schematic of an inline calcium looping CO_2 capture option using fluidised bed reactors (FB).

B.2 Integrated calcium looping option using fluidised bed reactors

The integrated fluidised bed calcium looping CO_2 capture configuration is somewhat less integrated into the clinker manufacturing process compared to the entrained flow solution presented in section 3.2. Nonetheless, it is characterised by a single oxy-fuel calciner calcining all raw meal. When fluidised bed reactors are employed a coarser milled raw meal fraction is required to run the dual fluidised bed system. Subsequently to the utilisation in the fluidised bed reactors the raw meal or calcium looping purge is milled to a finer particle size required for clinker burning. The calcium looping purge comprised all calcareous raw meal components required for clinker production. However, non-calcareous components such as silica sand or other correctives could be added to the second required milling step. The finely milled and pre-calcined raw meal mixture is then pre-heated in the cement plant's suspension pre-heater. Compared to the integrated entrained flow option no dedicated raw meal fraction is required to cool down the kilns flue gas before being fed to the carbonator since the carbonator is installed downstream of the cement plant's suspension pre-heater. The carbonator solely captures CO_2 from the combustion of the kiln as the CO_2 from raw meal calcination is already captured by oxy-fuel calcination. The large extent of purge cooling in this option, required due to the second milling step, will result in enhanced fuel consumption but will simultaneously increase the electricity output of the overall system.

B.3 Hybrid entrained flow fluidised bed calcium looping option

Besides the two elaborated reactor options a third hybrid option could be feasible combining the advantages of (i) entrained flow raw meal calcination at oxy-fuel conditions with (ii) the beneficial gas-solid contact of fluidised bed reactors while (iii) omitting the necessity to cool down large quantities of calcium looping purge for fine milling. The oxy-fuel calciner of the calcium looping system could be designed in such a way that operation with a bimodal raw meal mixture comprising a coarse and a fine fraction is possible. The coarse fraction will be separated from the CO_2 rich calciner flue gas in a first cyclone and utilised as fluidised bed calcium looping sorbent, whereas the finer fraction will solely be calcined and elutriated from the oxy-fuel calciner. In a second cyclone the fine particles are separated from the CO_2 rich flue gas and directed to the suspension pre-heater. Spent calcium looping sorbent will be purged from the calciner, cooled down and milled to a finesse required to form clinker in the rotary kiln and mixed with the finer particle fraction in the suspension pre-heater. Similar to the integrated fluidised bed configuration the carbonator only captures CO_2 from clinker burning since CO_2 from raw meal calcination is completely captured by oxy-fuel calcination in the hybrid entrained flow fluidised bed reactor. However, the fuel consumption is reduced due to the reduced purge (i.e. cooling and reheating). The effective make-up rate can be adjusted by means of the coarse raw meal share.

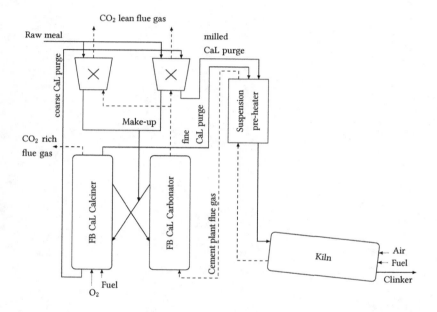

Figure B.3: Schematic of a calcium looping CO_2 capture option using CFB carbonator and a hybrid oxy-fuel calciner.

Bibliography

[1] ABANADES, C. J.: The maximum capture efficiency of CO2 using a carbonation/cal-cination cycle of CaO/CaCO3. *Chemical Engineering Journal* 90 (3), p. 303–306, 2002. DOI: http://dx.doi.org/10.1016/S1385-8947(02)00126-2

[2] ABANADES, C. J.; ALONSO, M.; RODRÍGUEZ, N.; GONZÁLEZ, B.; GRASA, G.; MURILLO, R.: Capturing CO2 from combustion flue gases with a carbonation calcination loop. Experimental results and process development. *Energy Procedia* 1 (1), p. 1147–1154, 2009. DOI: http://dx.doi.org/10.1016/j.egypro.2009.01.151

[3] ABANADES, C. J.; ALVAREZ, D.: Conversion Limits in the Reaction of CO2 with Lime. *Energy & Fuels* 17 (2), p. 308–315, 2003. DOI: http://dx.doi.org/10.1021/ef020152a

[4] ABANADES, C. J.; ANTHONY, E. J.; LU, D. Y.; SALVADOR, C.; ALVAREZ, D.: Capture of CO2 from combustion gases in a fluidized bed of CaO. *AIChE Journal* 50 (7), p. 1614–1622, 2004. DOI: http://dx.doi.org/10.1002/aic.10132

[5] ABANADES, C. J.; ANTHONY, E. J.; WANG, J.; OAKEY, J. E.: Fluidized Bed Combustion Systems Integrating CO2 Capture with CaO. *Environmental Science & Technology* 39 (8), p. 2861–2866, 2005. DOI: http://dx.doi.org/10.1021/es0496221

[6] ABANADES, C. J.; OAKEY, J. E.; ALVAREZ, D.; HÄMÄLÄINEN, J.: Novel Combustion Cycles Incorporating Capture of CO2 with CaO. In: GALE, J.; KAYA, Y.: Proceedings of the 6th International Conference on Greenhouse Gas Control Technologies, 1-4 October 2002, Kyoto, Japan, 1st ed.. Pergamon, 2003. ISBN: 978-0-08-044276-1. DOI: http://dx.doi.org/10.1016/B978-008044276-1/50029-5

[7] ALONSO, M.; ÁLVAREZ CRIADO, Y.; FERNÁNDEZ, J. R.; ABANADES, C.: CO2 Carrying Capacities of Cement Raw Meals in Calcium Looping Systems. *Energy & Fuels* 31 (12), p. 13955–13962, 2017. DOI: http://dx.doi.org/10.1021/acs.energyfuels.7b02586

© The Editor(s) (if applicable) and The Author(s), under exclusive license to
Springer Fachmedien Wiesbaden GmbH, part of Springer Nature 2022
M. Hornberger, *Experimental Investigation of Calcium Looping CO2
Capture for Application in Cement Plants*,
https://doi.org/10.1007/978-3-658-39248-2

[8] ALONSO, M.; ARIAS, B.; FERNÁNDEZ, J. R.; BUGHIN, O.; ABANADES, C.: Measuring attrition properties of calcium looping materials in a 30 kW pilot plant. *Powder Technology* 336, p. 273–281, 2018. DOI: http://dx.doi.org/10.1016/j.powtec.2018.06.011

[9] ALONSO, M.; CRIADO, Y. A.; ABANADES, C. J.; GRASA, G.: Undesired effects in the determination of CO_2 carrying capacities of CaO during TG testing. *Fuel* 127, p. 52–61, 2014. DOI: http://dx.doi.org/10.1016/j.fuel.2013.08.005

[10] ALONSO, M.; DIEGO, M. E.; PÉREZ, C.; CHAMBERLAIN, J. R.; ABANADES, C. J.: Biomass combustion with in situ CO_2 capture by CaO in a 300 kWth circulating fluidized bed facility. *International Journal of Greenhouse Gas Control* 29, p. 142–152, 2014. DOI: http://dx.doi.org/10.1016/j.ijggc.2014.08.002

[11] ALONSO, M.; FERNÁNDEZ, J. R.; ABANADES, C. J.: Kinetic Study of Belite Formation in Cement Raw Meals Used in the Calcium Looping CO_2 Capture Process. *Industrial & Engineering Chemistry Research* 58 (14), p. 5445–5454, 2019. DOI: http://dx.doi.org/1 0.1021/acs.iecr.9b00813

[12] ALONSO, M.; HORNBERGER, M.; SPÖRL, R.; SCHEFFKNECHT, G.; ABANADES, C.: Characterization of a Marl-Type Cement Raw Meal as CO_2 Sorbent for Calcium Looping. *ACS omega* 3 (11), p. 15229–15234, 2018. DOI: http://dx.doi.org/10.1021/acsomega.8b 01795

[13] ALONSO, M.; RODRÍGUEZ, N.; GRASA, G.; ABANADES, C. J.: Modelling of a fluidized bed carbonator reactor to capture CO_2 from a combustion flue gas. *Chemical Engineering Science* 64 (5), p. 883–891, 2009. DOI: http://dx.doi.org/10.1016/j.ces.2008.10.044

[14] ALVAREZ, D.; ABANADES, C. J.: Determination of the Critical Product Layer Thickness in the Reaction of CaO with CO_2. *Industrial & Engineering Chemistry Research* 44 (15), p. 5608–5615, 2005. DOI: http://dx.doi.org/10.1021/ie050305s

[15] ALVAREZ, D.; ABANADES, C. J.: Pore-Size and Shape Effects on the Recarbonation Performance of Calcium Oxide Submitted to Repeated Calcination/Recarbonation Cycles. *Energy & Fuels* 19 (1), p. 270–278, 2005. DOI: http://dx.doi.org/10.1021/e f049864m

[16] ALVAREZ, D.; PEÑA, M.; BORREGO, A. G.: Behavior of Different Calcium-Based Sorbents in a Calcination/Carbonation Cycle for CO_2 Capture. *Energy & Fuels* 21 (3), p. 1534–1542, 2007. DOI: http://dx.doi.org/10.1021/ef060573i

[17] ANANTHARAMAN, R.; BERSTAD, D.; CINTI, G.; LENA, E. de; GATTI, M.; HOPPE, H.; MARTINEZ, I.; MONTERIO, J. G. M.-S.; ROMANO, M.; ROUSSANALY, S.; SCHOLS, E.; SPINELLI, M.; STØRSET, S.; van OS, P.; VOLDSUND, M.: D3.2 CEMCAP framework for comparative techno-economic analysis of CO2 capture from cement plants - D3.2 (Version Revision 2). *Zenodo.* DOI: http://dx.doi.org/10.5281/zenodo.1257112

[18] ANTHONY, E. J.; GRANATSTEIN, D. L.: Sulfation phenomena in fluidized bed combustion systems. *Progress in Energy and Combustion Science* 27 (2), p. 215–236, 2001. DOI: http://dx.doi.org/10.1016/S0360-1285(00)00021-6

[19] ARIAS, B.; ALONSO, M.; ABANADES, C.: CO2 Capture by Calcium Looping at Relevant Conditions for Cement Plants Experimental Testing in a 30 kWth Pilot Plant. *Industrial & Engineering Chemistry Research* 56 (10), p. 2634–2640, 2017. DOI: http://dx.doi.org/10.1021/acs.iecr.6b04617

[20] ARIAS, B.; CORDERO, J. M.; ALONSO, M.; DIEGO, M. E.; ABANADES, C. J.: Investigation of SO2 Capture in a Circulating Fluidized Bed Carbonator of a Ca Looping Cycle. *Industrial & Engineering Chemistry Research* 52 (7), p. 2700–2706, 2013. DOI: http://dx.doi.org/10.1021/ie3026828

[21] ARIAS, B.; DIEGO, M. E.; MÉNDEZ, A.; ABANADES, C. J.; DÍAZ, L.; LORENZO, M.; SANCHEZ-BIEZMA, A.: Operating Experience in la Pereda 1.7 MWth Calcium Looping Pilot. *Energy Procedia* 114, p. 149–157, 2017. DOI: http://dx.doi.org/10.1016/j.egypro.2017.03.1157

[22] ARIAS, B.; DIEGO, M. E.; MÉNDEZ, A.; ALONSO, M.; ABANADES, C. J.: Calcium looping performance under extreme oxy-fuel combustion conditions in the calciner. *Fuel* 222, p. 711–717, 2018. DOI: http://dx.doi.org/10.1016/j.fuel.2018.02.163

[23] BALE, C. W.; BÉLISLE, E.; CHARTRAND, P.; DECTEROV, S. A.; ERIKSSON, G.; GHERIBI, A. E.; HACK, K.; JUNG, I.-H.; KANG, Y.-B.; MELANÇON, J.; PELTON, A. D.; PETERSEN, S.; ROBELIN, C.; SANGSTER, J.; SPENCER, P.; van ENDE, M.-A.: FactSage thermochemical software and databases, 2010–2016. *Calphad* 54, p. 35–53, 2016. DOI: http://dx.doi.org/10.1016/j.calphad.2016.05.002

[24] BARKER, R.: The reversibility of the reaction CaCO3 = CaO + CO2. *Journal of Applied Chemistry and Biotechnology* 23 (10), p. 733–742, 1973. DOI: http://dx.doi.org/10.1002/jctb.5020231005

[25] BARTHOLOMEW, C. H.: Sintering kinetics of supported metals: New perspectives from a unifying GPLE treatment. *Applied Catalysis A: General* 107 (1), p. 1–57, 1993. DOI: http://dx.doi.org/10.1016/0926-860X(93)85114-5

[26] BASU, P.: Combustion and gasification in fluidized beds. CRC/Taylor & Francis, 2006. ISBN: 9780849333965

[27] BEISHEIM, T.: Untersuchungen zum Verhalten von Kalkstein als Sorbens zur In-situ-Entschwefelung im Oxy-Fuel-Prozess mit zirkulierender Wirbelschichtfeuerung. Dissertation, University of Stuttgart, 2016. DOI: http://dx.doi.org/10.18419/opus -8804

[28] BERRUTI, F.; PUGSLEY, T. S.; GODFROY, L.; CHAOUKI, J.; PATIENCE, G. S.: Hydrodynamics of circulating fluidized bed risers: A review. *The Canadian Journal of Chemical Engineering* 73 (5), p. 579–602, 1995. DOI: http://dx.doi.org/10.1002/cjce.5450730502

[29] BHATIA, S. K.; PERLMUTTER, D. D.: Effect of the product layer on the kinetics of the CO_2-lime reaction. *AIChE Journal* 29 (1), p. 79–86, 1983. DOI: http://dx.doi.org/10. 1002/aic.690290111

[30] BLAMEY, J.; ANTHONY, E. J.; WANG, J.; FENNELL, P. S.: The calcium looping cycle for large-scale CO_2 capture. *Progress in Energy and Combustion Science* 36 (2), p. 260–279, 2010. DOI: http://dx.doi.org/10.1016/j.pecs.2009.10.001

[31] BOLDT, M.: CO2-Abscheidung in der Zementindustrie mittels Calcium Looping – feststoffseitige Bilanzierung. Bachelor thesis, University of Stuttgart, 2017

[32] BORGWARDT, R. H.: Calcination kinetics and surface area of dispersed limestone particles. *AIChE Journal* 31 (1), p. 103–111, 1985. DOI: http://dx.doi.org/10.1002/aic .690310112

[33] BORGWARDT, R. H.: Calcium oxide sintering in atmospheres containing water and carbon dioxide. *Industrial & Engineering Chemistry Research* 28 (4), p. 493–500, 1989. DOI: http://dx.doi.org/10.1021/ie00088a019

[34] BORGWARDT, R. H.: Sintering of nascent calcium oxide. *Chemical Engineering Science* 44 (1), p. 53–60, 1989. DOI: http://dx.doi.org/10.1016/0009-2509(89)85232-7

[35] BOUQUET, E.; LEYSSENS, G.; SCHÖNNENBECK, C.; GILOT, P.: The decrease of car-bonation efficiency of CaO along calcination–carbonation cycles: Experiments and modelling. *Chemical Engineering Science* 64 (9), p. 2136–2146, 2009. DOI: http://dx.doi.org/10.1016/j.ces.2009.01.045

[36] BRAND, T.: Experimentelle Untersuchung zur Charakterisierung von Rohmehl als Calcium-Looping-Sorbens in Flugstromreaktoren – Sorbenskalzinierung und Rekar-bonatisierung. Bachelor thesis, University of Stuttgart, 2018

[37] CHARITOS, A.: Experimental Characterization of the Calcium Looping Process for CO2 Capture. Dissertation, University of Stuttgart, 2013. DOI: http://dx.doi.org/10.18419/opus-2174

[38] CHARITOS, A.; HAWTHORNE, C.; BIDWE, A. R.; SIVALINGAM, S.; SCHUSTER, A.; SPLI-ETHOFF, H.; SCHEFFKNECHT, G.: Parametric investigation of the calcium looping process for CO2 capture in a 10kWth dual fluidized bed. *International Journal of Greenhouse Gas Control* 4 (5), p. 776–784, 2010. DOI: http://dx.doi.org/10.1016/j.ijggc.2010.04.009

[39] CHARITOS, A.; RODRÍGUEZ, N.; HAWTHORNE, C.; ALONSO, M.; ZIEBA, M.; ARIAS, B.; KOPANAKIS, G.; SCHEFFKNECHT, G.; ABANADES, C. J.: Experimental Validation of the Calcium Looping CO2 Capture Process with Two Circulating Fluidized Bed Carbona-tor Reactors. *Industrial & Engineering Chemistry Research* 50 (16), p. 9685–9695, 2011. DOI: http://dx.doi.org/10.1021/ie200579f

[40] CHEN, C.; ZHAO, C.; LIANG, C.; PANG, K.: Calcination and sintering characteristics of limestone under O2/CO2 combustion atmosphere. *Fuel Processing Technology* 88 (2), p. 171–178, 2007. DOI: http://dx.doi.org/10.1016/j.fuproc.2006.03.003

[41] CINTI, G.; LENA, E. de; SPINELLI, M.; ROMANO, M.: D12.5 Design of post combustion capture and integrated calcium looping cement plant systems. *Zenodo* 2018. DOI: http://dx.doi.org/10.5281/zenodo.2605104

[42] COPPOLA, A.; ESPOSITO, A.; MONTAGNARO, F.; IULIANO, M.; SCALA, F.; SALATINO, P.: The combined effect of H2O and SO2 on CO2 uptake and sorbent attrition dur-ing fluidised bed calcium looping. *Proceedings of the Combustion Institute* 37 (4), p. 4379–4387, 2019. DOI: http://dx.doi.org/10.1016/j.proci.2018.08.013

[43] COPPOLA, A.; SCALA, F.; SALATINO, P.; MONTAGNARO, F.: Fluidized bed calcium loop-
     ing cycles for CO2 capture under oxy-firing calcination conditions: Part 1. Assess-
     ment of six limestones. *Chemical Engineering Journal* 231, p. 537–543, 2013. DOI:
     http://dx.doi.org/10.1016/j.cej.2013.07.113

[44] CORRENS, C. W.: Einführung in die Mineralogie Kristallographie und Petrologie,
     Zweite Auflage Springer Berlin Heidelberg, 1968. ISBN: 978-3-662-23826-4. DOI: http:
     //dx.doi.org/10.1007/978-3-662-25929-0

[45] CRIADO, Y. A.; ARIAS, B.; ABANADES, C. J.: Effect of the Carbonation Temperature on
     the CO2 Carrying Capacity of CaO. *Industrial & Engineering Chemistry Research* 57
     (37), p. 12595–12599, 2018. DOI: http://dx.doi.org/10.1021/acs.iecr.8b02111

[46] CURRAN, G. P.; FINK, C. E.; GORIN, E.:   CO2 Acceptor Gasification Process. In:
     SCHORA, F. C.: Fuel Gasification. American Chemical Society, 1967. ISBN: 0-8412-
     0070-X. DOI: http://dx.doi.org/10.1021/ba-1967-0069.ch010

[47] DARROUDI, T.; SEARCY, A. W.: Effect of carbon dioxide pressure on the rate of decom-
     position of calcite (CaCO3). *The Journal of Physical Chemistry* 85 (26), p. 3971–3974,
     1981. DOI: http://dx.doi.org/10.1021/j150626a004

[48] DENNIS, J. S.; HAYHURST, A. N.: The effect of CO2 on the kinetics and extent of calci-
     nation of limestone and dolomite particles in fluidised beds. *Chemical Engineering Sci-
     ence* 42 (10), p. 2361–2372, 1987. DOI: http://dx.doi.org/10.1016/0009-2509(87)80110-0

[49] DIEGO, L. F.; RUFAS, A.; GARCÍA-LABIANO, F.; LAS OBRAS-LOSCERTALES, M. de; ABAD,
     A.; GAYÁN, P.; ADÁNEZ, J.: Optimum temperature for sulphur retention in fluidised
     beds working under oxy-fuel combustion conditions. *Fuel* 114, p. 106–113, 2013. DOI:
     http://dx.doi.org/10.1016/j.fuel.2012.02.064

[50] DIEGO, M. E.; ALONSO, M.: Operational feasibility of biomass combustion with in
     situ CO2 capture by CaO during 360 h in a 300 kWth calcium looping facility. *Fuel*
     181, p. 325–329, 2016. DOI: http://dx.doi.org/10.1016/j.fuel.2016.04.128

[51] DIEGO, M. E.; ARIAS, B.: Impact of load changes on the carbonator reactor of a 1.7
     MWth calcium looping pilot plant. *Fuel Processing Technology* 200, p. 106307, 2020.
     DOI: http://dx.doi.org/10.1016/j.fuproc.2019.106307

[52] DIEGO, M. E.; ARIAS, B.; MÉNDEZ, A.; LORENZO, M.; DÍAZ, L.; SÁNCHEZ-BIEZMA, A.; ABANADES, C. J.: Experimental testing of a sorbent reactivation process in La Pereda 1.7 MWth calcium looping pilot plant. *International Journal of Greenhouse Gas Control* 50, p. 14–22, 2016. DOI: http://dx.doi.org/10.1016/j.ijggc.2016.04.008

[53] DIEGO, M. E.; MARTÍNEZ, I.; ALONSO, M.; ARIAS, B.; ABANADES, C. J.: Calcium looping reactor design for fluidized-bed systems: Calcium and Chemical Looping Technology for Power Generation and Carbon Dioxide (CO2) Capture. Elsevier, 2015. ISBN: 9780857092434. DOI: http://dx.doi.org/10.1016/B978-0-85709-243-4.00006-9

[54] DIETER, H.; BIDWE, A. R.; VARELA-DUELLI, G.; CHARITOS, A.; HAWTHORNE, C.; SCHEFF-KNECHT, G.: Development of the calcium looping CO2 capture technology from lab to pilot scale at IFK, University of Stuttgart. *Fuel* 127, p. 23–37, 2014. DOI: http://dx.doi.org/10.1016/j.fuel.2014.01.063

[55] DIETER, H.; HAWTHORNE, C.; ZIEBA, M.; SCHEFFKNECHT, G.: Progress in Calcium Looping Post Combustion CO2 Capture: Successful Pilot Scale Demonstration. *Energy Procedia* 37, p. 48–56, 2013. DOI: http://dx.doi.org/10.1016/j.egypro.2013.05.084

[56] DOBNER, S.; STERNS, L.; GRAFF, R. A.; SQUIRES, A. M.: Cyclic Calcination and Recarbonation of Calcined Dolomite. *Industrial & Engineering Chemistry Process Design and Development* 16 (4), p. 479–486, 1977. DOI: http://dx.doi.org/10.1021/i260064a008

[57] DONAT, F.; FLORIN, N. H.; ANTHONY, E. J.; FENNELL, P. S.: Influence of high-temperature steam on the reactivity of CaO sorbent for CO_2 capture. *Environmental Science & Technology* 46 (2), p. 1262–1269, 2012. DOI: http://dx.doi.org/10.1021/es2 02679w

[58] DOU, B.; SONG, Y.; LIU, Y.; FENG, C.: High temperature CO2 capture using calcium oxide sorbent in a fixed-bed reactor. *Journal of hazardous materials* 183 (1-3), p. 759–765, 2010. DOI: http://dx.doi.org/10.1016/j.jhazmat.2010.07.091

[59] DUELLI, G.; BIDWE, A. R.; PAPANDREOU, I.; DIETER, H.; SCHEFFKNECHT, G.: Characterization of the oxy-fired regenerator at a 10 kWth dual fluidized bed calcium looping facility. *Applied Thermal Engineering* 74, p. 54–60, 2015. DOI: http://dx.doi.org/10.10 16/j.applthermaleng.2014.03.042

[60] DUELLI, G.; CHARITOS, A.; DIEGO, M. E.; STAVROULAKIS, E.; DIETER, H.; SCHEF-
FKNECHT, G.: Investigations at a 10 kWth calcium looping dual fluidized bed facility Limestone calcination and CO2 capture under high CO2 and water vapor atmosphere. *International Journal of Greenhouse Gas Control* 33, p. 103–112, 2015. DOI:
http://dx.doi.org/10.1016/j.ijggc.2014.12.006

[61] ELLIS, L. D.; BADEL, A. F.; CHIANG, M. L.; PARK, R. J.-Y.; CHIANG, Y.-M.: Toward
electrochemical synthesis of cement-An electrolyzer-based process for decarbonat-
ing CaCO3 while producing useful gas streams. *Proceedings of the National Academy
of Sciences of the United States of America* 117 (23), p. 12584–12591, 2020. DOI:
http://dx.doi.org/10.1073/pnas.1821673116

[62] FANG, F.; LI, Z.-S.; CAI, N.-S.: Experiment and Modeling of CO2 Capture from Flue
Gases at High Temperature in a Fluidized Bed Reactor with Ca-Based Sorbents. *Energy & Fuels* 23 (1), p. 207–216, 2009. DOI: http://dx.doi.org/10.1021/ef800474n

[63] FENNELL, P.: Economics of chemical and calcium looping: Calcium and Chemical
Looping Technology for Power Generation and Carbon Dioxide (CO2) Capture. Elsevier, 2015. ISBN: 9780857092434. DOI: http://dx.doi.org/10.1016/B978-0-85709-243
-4.00003-3

[64] FENNELL, P. S.; PACCIANI, R.; DENNIS, J. S.; DAVIDSON, J. F.; HAYHURST, A. N.: The
Effects of Repeated Cycles of Calcination and Carbonation on a Variety of Different
Limestones, as Measured in a Hot Fluidized Bed of Sand. *Energy & Fuels* 21 (4), p.
2072–2081, 2007. DOI: http://dx.doi.org/10.1021/ef060506o

[65] FERNÁNDEZ, J. R.; ALONSO, M.; TURRADO, S.; ARIAS, B.; ABANADES, C.: D4.6 Experimental determination of intrinsic calcination/carbonation kinetics for various raw
meals. CLEANKER Consortium, 2019

[66] FERNANDEZ, J. R.; TURRADO, S.; ABANADES, C. J.: Calcination kinetics of cement raw
meals under various CO2 concentrations. *Reaction Chemistry & Engineering* 4 (12),
p. 2129–2140, 2019. DOI: http://dx.doi.org/10.1039/C9RE00361D

[67] FUERTES, A. B.; ALVAREZ, D.; RUBIERA, F.; PIS, J. J.; MARBÁN, G.; PALACIOS, J. M.: Surface area and pore size changes during sintering of calcium oxide particles. *Chemical
Engineering Communications* 109 (1), p. 73–88, 1990. DOI: http://dx.doi.org/10.1080
/00986449108910974

[68] GARCÍA-LABIANO, F.; ABAD, A.; DIEGO, L. F.; GAYÁN, P.; ADÁNEZ, J.: Calcination of calcium-based sorbents at pressure in a broad range of CO2 concentrations. *Chemical Engineering Science* 57 (13), p. 2381–2393, 2002. DOI: http://dx.doi.org/10.1016/S0009 -2509(02)00137-9

[69] GARCÍA-LABIANO, F.; RUFAS, A.; DIEGO, L. F.; OBRAS-LOSCERTALES, M. d. l.; GAYÁN, P.; ABAD, A.; ADÁNEZ, J.: Calcium-based sorbents behaviour during sulphation at oxy-fuel fluidised bed combustion conditions. *Fuel* 90 (10), p. 3100–3108, 2011. DOI: http://dx.doi.org/10.1016/j.fuel.2011.05.001

[70] GELDART, D.: Types of gas fluidization. *Powder Technology* 7 (5), p. 285–292, 1973. DOI: http://dx.doi.org/10.1016/0032-5910(73)80037-3

[71] GERMAN INSTITUTE FOR STANDARDIZATION: DIN EN 14792:2017-05, Emissionen aus stationären Quellen- Bestimmung der Massenkonzentration von Stickstoffoxiden - Standardreferenzverfahren: Chemilumineszenz; Deutsche Fassung EN 14792:2017, 2017. DOI: http://dx.doi.org/10.31030/2463423

[72] GERMAN INSTITUTE FOR STANDARDIZATION: DIN EN 197-1:2018-11, Zement - Teil 1: Zusammensetzung, Anforderungen und Konformitätskriterien von Normalzement; Deutsche und Englische Fassung prEN 197-1:2018, 2018. DOI: http://dx.doi.org/10. 31030/2874267

[73] GONZÁLEZ, B.; GRASA, G. S.; ALONSO, M.; ABANADES, C. J.: Modeling of the De-activation of CaO in a Carbonate Loop at High Temperatures of Calcination. *Industrial & Engineering Chemistry Research* 47 (23), p. 9256–9262, 2008. DOI: http://dx.doi.org/10.1021/ie8009318

[74] GRASA, G.; ABANADES, C. J.; ALONSO, M.; GONZÁLEZ, B.: Reactivity of highly cycled particles of CaO in a carbonation/calcination loop. *Chemical Engineering Journal* 137 (3), p. 561–567, 2008. DOI: http://dx.doi.org/10.1016/j.cej.2007.05.017

[75] GRASA, G.; MURILLO, R.; ALONSO, M.; ABANADES, C. J.: Application of the random pore model to the carbonation cyclic reaction. *AIChE Journal* 55 (5), p. 1246–1255, 2009. DOI: http://dx.doi.org/10.1002/aic.11746

[76] GRASA, G. S.; ABANADES, C. J.: CO2 Capture Capacity of CaO in Long Series of Carbonation/Calcination Cycles. *Industrial & Engineering Chemistry Research* 45 (26), p. 8846–8851, 2006. DOI: http://dx.doi.org/10.1021/ie0606946

[77] GRASA, G. S.; ALONSO, M.; ABANADES, C. J.: Sulfation of CaO Particles in a Carbonation/Calcination Loop to Capture CO2. *Industrial & Engineering Chemistry Research* 47 (5), p. 1630–1635, 2008. DOI: http://dx.doi.org/10.1021/ie070937%2B

[78] GUPTA, H.; FAN, L.-S.: Carbonation–Calcination Cycle Using High Reactivity Calcium Oxide for Carbon Dioxide Separation from Flue Gas. *Industrial & Engineering Chemistry Research* 41 (16), p. 4035–4042, 2002. DOI: http://dx.doi.org/10.1021/ie010 8671

[79] GUPTA, S. K.; BERRUTI, F.: Evaluation of the gas–solid suspension density in CFB risers with exit effects. *Powder Technology* 108 (1), p. 21–31, 2000. DOI: http://dx.doi .org/10.1016/S0032-5910(99)00199-0

[80] HAWTHORNE C.; CHARITOS, A.; PEREZ-PULIDO C.A.; BING Z.; SCHEFFKNECHT, G.: Design of a dual fluidised bed system for the post combustion removal of CO2 using CaO. Part 1: CFB Carbonator reactor model. In: WERTHER, J.: Proceedings of the 9th International Conference on Circulating Fluidized Beds. TuTech Innovation GmbH, 2008. ISBN: 393040057X

[81] HILZ, J.; HELBIG, M.; HAAF, M.; DAIKELER, A.; STRÖHLE, J.; EPPLE, B.: Long-term pilot testing of the carbonate looping process in 1 MWth scale. *Fuel* 210, p. 892–899, 2017. DOI: http://dx.doi.org/10.1016/j.fuel.2017.08.105

[82] HILZ, J.; HELBIG, M.; HAAF, M.; DAIKELER, A.; STRÖHLE, J.; EPPLE, B.: Investigation of the fuel influence on the carbonate looping process in 1 MWth scale. *Fuel Processing Technology* 169, p. 170–177, 2018. DOI: http://dx.doi.org/10.1016/j.fuproc.2017.09.016

[83] HOFBAUER, G.: Experimentelle Untersuchung der Oxy-Fuel-Verbrennung von Steinkohle in einer zirkulierenden Wirbelschichtfeuerung. Dissertation, University of Stuttgart, 2017. DOI: http://dx.doi.org/10.18419/opus-9129

[84] HORNBERGER, M.: Characterization of cement raw meal as CO2 sorbent in an entrained flow calcium looping CO2 capture system for cement plants. In: High Temperature Solid Looping Cycle Network Meeting, 20-21 Jan. 2020, Geleen, The Netherlands

[85] HORNBERGER, M.; MORENO, J.; SCHMID, M.; SCHEFFKNECHT, G.: Experimental investigation of the carbonation reactor in a tail-end Calcium Looping configuration for CO2 capture from cement plants. *Fuel Processing Technology* 210, p. 106557, 2020. DOI: http://dx.doi.org/10.1016/j.fuproc.2020.106557

[86] HORNBERGER, M.; MORENO, J.; SCHMID, M.; SCHEFFKNECHT, G.: Experimental investigation of the calcination reactor in a tail-end calcium looping configuration for CO2 capture from cement plants. *Fuel* 284, p. 118927, 2021. DOI: http://dx.doi.org/1 0.1016/j.fuel.2020.118927

[87] HORNBERGER, M.; SPÖRL, R.; SCHEFFKNECHT, G.: Calcium Looping for CO2 capture in the cement industry – pilot scale experiments. In: MIREK, P.; NOWAK, W.; ŚCIĄŻKO, M.: Proceedings of the 12th International Conference of Fluidized Bed Technology. Fundacja dla Akademii Górniczo-Hutniczej im. Stanisława Staszica, 2017. ISBN: 978-83-62079-16-2

[88] HORNBERGER, M.; SPÖRL, R.; SCHEFFKNECHT, G.: Demonstration of Calcium Looping CO2 Capture for Cement Plants at Semi Industrial Scale. *Proceedings of the 14th Greenhouse Gas Control Technologies Conference Melbourne 21-26 October 2018 (GHGT-14).* DOI: http://dx.doi.org/10.2139/ssrn.3366038

[89] HORNBERGER, M.; SPÖRL, R.; SCHEFFKNECHT, G.: Pilot scale demonstration of CO2 capture by Calcium Looping under cement plant specific conditions. In: Proceedings of the 23rd International Conference on Fluidized Bed Conversion, 13-17 May 2018, Seoul, Korea

[90] HORNBERGER, M.; SPÖRL, R.; SCHEFFKNECHT, G.: Calcium Looping for CO2 Capture in Cement Plants – Pilot Scale Test. *Energy Procedia* 114, p. 6171–6174, 2017. DOI: http://dx.doi.org/10.1016/j.egypro.2017.03.1754

[91] HORNBERGER, M.; SPÖRL, R.; SCHEFFKNECHT, G.: CCS in cement industry – Application of the Calcium Looping Technology. In: The Trondheim Conference on Carbon Capture, Transport and Storage, 12-14 June 2017, Trondheim, Norway

[92] HORNBERGER, M.; SPÖRL, R.; SCHEFFKNECHT, G.: Experimental investigation on emission-free cement production by Calcium Looping post combustion CO2 capture. In: High Temperature Solid Looping Cycle Network Meeting, 04-05 Sept. 2017, Luleå, Sweden

[93] HU, N.; SCARONI, A. W.: Calcination of pulverized limestone particles under furnace injection conditions. *Fuel* 75 (2), p. 177–186, 1996. DOI: http://dx.doi.org/10.1016/0 016-2361(95)00234-0

[94] HUGHES, R. W.; LU, D.; ANTHONY, E. J.; WU, Y.: Improved Long-Term Conversion of Limestone-Derived Sorbents for In Situ Capture of CO2 in a Fluidized Bed Combustor. *Industrial & Engineering Chemistry Research* 43 (18), p. 5529–5539, 2004. DOI: http://dx.doi.org/10.1021/ie034260b

[95] IEA: Technology Roadmap - Low-Carbon Transition in the Cement Industry. IEA, Paris, 2018. https://www.iea.org/reports/technology-roadmap-low-carbon-transitio n-in-the-cement-industry (Access on 01.08.2019)

[96] INGRAHAM, T. R.; MARIER, P.: Kinetic studies on the thermal decomposition of calcium carbonate. *The Canadian Journal of Chemical Engineering* 41 (4), p. 170–173, 1963. DOI: http://dx.doi.org/10.1002/cjce.5450410408

[97] IPCC 2018: Global Warming of 1.5 °C. An IPCC Special Report on the impacts of global warming of 1.5 °C above pre-industrial levels and related global greenhouse gas emission pathways, in the context of strengthening the global response to the threat of climate change, sustainable development, and efforts to eradicate poverty. Masson-Delmotte, V. and Zhai, P. and Pörtner, H.-O. and Roberts, D. and Skea, J. and Shukla, P. R. and Pirani, A. and Moufouma-Okia, W. and Péan, C. and Pidcock, R. and Connors, S. and Matthews, J.B.R. and Chen, Y. and Zhou, X. and Gomis, M. I. and Lonnoy, E. and Maycock, T. and Tignor, M. and Waterfield, T., Geneva, 2018

[98] JOHNSSON, J. E.: Formation and reduction of nitrogen oxides in fluidized-bed combustion. *Fuel* 73 (9), p. 1398–1415, 1994. DOI: http://dx.doi.org/10.1016/0016-2361(94)90055-8

[99] JUNK, M.; REITZ, M.; STRÖHLE, J.; EPPLE, B.: Technical and Economical Assessment of the Indirectly Heated Carbonate Looping Process. *Journal of Energy Resources Technology* 138 (4), p. 1051, 2016. DOI: http://dx.doi.org/10.1115/1.4033142

[100] KEIL, F.: Zement Herstellung und Eigenschaften. Springer Berlin Heidelberg, 1971. ISBN: 978-3-642-80578-3. DOI: http://dx.doi.org/10.1007/978-3-642-80577-6

[101] KIRSCH, J.: Sorbenscharakterisierung zur Flugstrom Calcium Looping CO2 Abscheidung aus der Zementindustrie. Master thesis, University of Stuttgart, 2021

[102] KUNII, D.; LEVENSPIEL, O.; BRENNER, H.: Fluidization Engineering, 2nd ed. Elsevier Science, 2013. ISBN: 978-0-08-050664-7

[103] LAS OBRAS-LOSCERTALES, M. de; RUFAS, A.; DIEGO, L. F.; GARCÍA-LABIANO, F.; GAYÁN, P.; ABAD, A.; ADÁNEZ, J.: Effects of Temperature and Flue Gas Recycle on the SO2 and NOx Emissions in an Oxy-fuel Fluidized Bed Combustor. Energy Procedia 37, p. 1275–1282, 2013. DOI: http://dx.doi.org/10.1016/j.egypro.2013.06.002

[104] LASHERAS, A.; STRÖHLE, J.; GALLOY, A.; EPPLE, B.: Carbonate looping process simulation using a 1D fluidized bed model for the carbonator. International Journal of Greenhouse Gas Control 5 (4), p. 686–693, 2011. DOI: http://dx.doi.org/10.1016/j.ijggc .2011.01.005

[105] LENA, E. de; SPINELLI, M.; MARTÍNEZ, I.; GATTI, M.; SCACCABAROZZI, R.; CINTI, G.; ROMANO, M. C.: Process integration study of tail-end Ca-Looping process for CO2 capture in cement plants. International Journal of Greenhouse Gas Control 67, p. 71–92, 2017. DOI: http://dx.doi.org/10.1016/j.ijggc.2017.10.005

[106] LENA, E. d.; SPINELLI, M.; GATTI, M.; SCACCABAROZZI, R.; CAMPANARI, S.; CONSONNI, S.; CINTI, G.; ROMANO, M. C.: Techno-economic analysis of calcium looping processes for low CO2 emission cement plants. International Journal of Greenhouse Gas Control 82, p. 244–260, 2019. DOI: http://dx.doi.org/10.1016/j.ijggc.2019.01.005

[107] LI, H.; YAN, J.; ANHEDEN, M.: Impurity impacts on the purification process in oxy-fuel combustion based CO2 capture and storage system. Applied Energy 86 (2), p. 202–213, 2009. DOI: http://dx.doi.org/10.1016/j.apenergy.2008.05.006

[108] LI, S.; XU, M.; JIA, L.; TAN, L.; LU, Q.: Influence of operating parameters on N2O emission in O2/CO2 combustion with high oxygen concentration in circulating fluidized bed. Applied Energy 173, p. 197–209, 2016. DOI: http://dx.doi.org/10.1016/j.apenerg y.2016.02.054

[109] LI, Z.; SUN, H.; CAI, N.: Rate Equation Theory for the Carbonation Reaction of CaO with CO2. Energy & Fuels 26 (7), p. 4607–4616, 2012. DOI: http://dx.doi.org/10.1021 /ef300607z

[110] LI, Z.-S.; FANG, F.; TANG, X.-y.; CAI, N.-S.: Effect of Temperature on the Carbonation Reaction of CaO with CO2. Energy & Fuels 26 (4), p. 2473–2482, 2012. DOI: http: //dx.doi.org/10.1021/ef201543n

[111] LIU, H.; GIBBS, B.: The influence of calcined limestone on NOx and N2O emissions from char combustion in fluidized bed combustors. *Fuel* 80 (9), p. 1211–1215, 2001. DOI: http://dx.doi.org/10.1016/S0016-2361(00)00212-X

[112] LIU, W.; DENNIS, J. S.; SULTAN, D. S.; REDFERN, S. A.; SCOTT, S. A.: An investigation of the kinetics of CO2 uptake by a synthetic calcium based sorbent. *Chemical Engineering Science* 69 (1), p. 644–658, 2012. DOI: http://dx.doi.org/10.1016/j.ces.2011.11.036

[113] LOCHER, F.: Zement Grundlagen der Herstellung und Verwendung, 1. Aufl. Verlag Bau+Technik, 2015. ISBN: 3-7640-0400-2

[114] LOCHER, F. W.: Cement Principles of production and use. Verl. Bau + Technik, 2006. ISBN: 3764004207

[115] LOCKWOOD, F.; GRANADOS, L.; LECLERC, M.; LESORT, A.-L.; BEASSE, G.; DELGADO, M. A.; SPERO, C.: Oxy-combustion CPU – From Pilots Towards Industrial-scale Demonstration. *Energy Procedia* 63, p. 342–351, 2014. DOI: http://dx.doi.org/10. 1016/j.egypro.2014.11.037

[116] LU, D. Y.; HUGHES, R. W.; ANTHONY, E. J.: Ca-based sorbent looping combustion for CO2 capture in pilot-scale dual fluidized beds. *Fuel Processing Technology* 89 (12), p. 1386–1395, 2008. DOI: http://dx.doi.org/10.1016/j.fuproc.2008.06.011

[117] MAKOSKI, P.: Vergleichende Untersuchung zur Eignung von Rohmehlen als Calcium Looping Sorbens. Bachelor thesis, University of Stuttgart, 2019

[118] MANOVIC, V.; ANTHONY, E. J.: Parametric Study on the CO2 Capture Capacity of CaO-Based Sorbents in Looping Cycles. *Energy & Fuels* 22 (3), p. 1851–1857, 2008. DOI: http://dx.doi.org/10.1021/ef800011z

[119] MANOVIC, V.; ANTHONY, E. J.: Sequential SO2/CO2 capture enhanced by steam reactivation of a CaO-based sorbent. *Fuel* 87 (8-9), p. 1564–1573, 2008. DOI: http: //dx.doi.org/10.1016/j.fuel.2007.08.022

[120] MANOVIC, V.; ANTHONY, E. J.: Carbonation of CaO-Based Sorbents Enhanced by Steam Addition. *Industrial & Engineering Chemistry Research* 49 (19), p. 9105–9110, 2010. DOI: http://dx.doi.org/10.1021/ie101352s

[121] MANOVIC, V.; ANTHONY, E. J.; LONCAREVIC, D.: SO2 Retention by CaO-Based Sorbent Spent in CO2 Looping Cycles. *Industrial & Engineering Chemistry Research* 48 (14), p. 6627–6632, 2009. DOI: http://dx.doi.org/10.1021/ie9002365

[122] MANOVIC, V.; CHARLAND, J.-P.; BLAMEY, J.; FENNELL, P. S.; LU, D. Y.; ANTHONY, E. J.: Influence of calcination conditions on carrying capacity of CaO-based sorbent in CO2 looping cycles. *Fuel* 88 (10), p. 1893–1900, 2009. DOI: http://dx.doi.org/10.1016/j.fue l.2009.04.012

[123] MARTÍNEZ, I.; GRASA, G.; MURILLO, R.; ARIAS, B.; ABANADES, C. J.: Kinetics of Calcination of Partially Carbonated Particles in a Ca-Looping System for CO2 Capture. *Energy & Fuels* 26 (2), p. 1432–1440, 2012. DOI: http://dx.doi.org/10.1021/ef201525k

[124] MARTÍNEZ, I.; GRASA, G.; MURILLO, R.; ARIAS, B.; ABANADES, C. J.: Modelling the continuous calcination of CaCO3 in a Ca-looping system. *Chemical Engineering Journal* 215-216, p. 174–181, 2013. DOI: http://dx.doi.org/10.1016/j.cej.2012.09.134

[125] MARTÍNEZ, I.; GRASA, G.; PARKKINEN, J.; TYNJÄLÄ, T.; HYPPÄNEN, T.; MURILLO, R.; ROMANO, M. C.: Review and research needs of Ca-Looping systems modelling for post-combustion CO2 capture applications. *International Journal of Greenhouse Gas Control* 50, p. 271–304, 2016. DOI: http://dx.doi.org/10.1016/j.ijggc.2016.04.002

[126] MICCIO, F.; LÖFFLER, G.; WARGADALAM, V. J.; WINTER, F.: The influence of SO2 level and operating conditions on NOx and N2O emissions during fluidised bed combustion of coals. *Fuel* 80 (11), p. 1555–1566, 2001. DOI: http://dx.doi.org/10.1016/S0016 -2361(01)00029-1

[127] MORENO, J.; HORNBERGER, M.; SCHMID, M.; SCHEFFKNECHT, G.: Part-Load Operation of a Novel Calcium Looping System for Flexible CO2 Capture in Coal-Fired Power Plants. *Industrial & Engineering Chemistry Research*, 2021. DOI: http://dx.doi.org/10. 1021/acs.iecr.1c00155

[128] MORENO, J.; SPÖRL, R.; SCHEFFKNECHT, G.: Increased load flexibility of the Calcium Looping CO2 capture process with a bubbling/turbulent fluidized bed carbonator. In: Proceeding of the 23rd International Conference on Fluidized Bed Conversion, 13-17 May 2018, Seoul, Korea

[129] MORENO, J.; SPÖRL, R.; SCHEFFKNECHT, G.: CO2 Capture with a Highly Flexible Calcium Looping System Using a BFB/TFB Carbonator. *SSRN Electronic Journal*, 2019. DOI: http://dx.doi.org/10.2139/ssrn.3332625

[130] NIKOLOPOULOS, A.; NIKOLOPOULOS, N.; CHARITOS, A.; GRAMMELIS, P.; KAKARAS,
       E.; BIDWE, A. R.; VARELA, G.: High-resolution 3-D full-loop simulation of a CFB
       carbonator cold model. *Chemical Engineering Science* 90, p. 137–150, 2013. DOI:
       http://dx.doi.org/10.1016/j.ces.2012.12.007

[131] NIKOLOPOULOS, A.; STROH, A.; ZENELI, M.; ALOBAID, F.; NIKOLOPOULOS, N.; STRÖHLE,
       J.; KARELLAS, S.; EPPLE, B.; GRAMMELIS, P.: Numerical investigation and comparison
       of coarse grain CFD – DEM and TFM in the case of a 1 MWth fluidized bed carbonator
       simulation. *Chemical Engineering Science* 163, p. 189–205, 2017. DOI: http://dx.doi.o
       rg/10.1016/j.ces.2017.01.052

[132] OKUNEV, A. G.; NESTERENKO, S. S.; LYSIKOV, A. I.: Decarbonation Rates of Cycled
       CaO Absorbents. *Energy & Fuels* 22 (3), p. 1911–1916, 2008. DOI: http://dx.doi.org/1
       0.1021/ef800047b

[133] ORTIZ, C.; VALVERDE, J. M.; CHACARTEGUI, R.; BENÍTEZ-GUERRERO, M.; PEREJÓN, A.;
       ROMEO, L. M.: The Oxy-CaL process: A novel CO2 capture system by integrating
       partial oxy-combustion with the Calcium-Looping process. *Applied Energy* 196, p.
       1–17, 2017. DOI: http://dx.doi.org/10.1016/j.apenergy.2017.03.120

[134] PATHI, S. K.; LIN, W.; ILLERUP, J. B.; DAM-JOHANSEN, K.; HJULER, K.: CO2 Capture by
       Cement Raw Meal. *Energy & Fuels* 27 (9), p. 5397–5406, 2013. DOI: http://dx.doi.org
       /10.1021/ef401073p

[135] PEREJÓN, A.; ROMEO, L. M.; LARA, Y.; LISBONA, P.; MARTÍNEZ, A.; VALVERDE, J. M.:
       The Calcium-Looping technology for CO2 capture On the important roles of energy
       integration and sorbent behavior. *Applied Energy* 162, p. 787–807, 2016. DOI: http:
       //dx.doi.org/10.1016/j.apenergy.2015.10.121

[136] RAO, R. T.; GUNN, D. J.; BOWEN, J. H.: Kinetics of calcium carbonate decomposition.
       Chemical engineering research & design 67 (1), p. 38–47, 1989

[137] RODRÍGUEZ, N.; ALONSO, M.; ABANADES, C. J.: Average activity of CaO particles in a
       calcium looping system. *Chemical Engineering Journal* 156 (2), p. 388–394, 2010. DOI:
       http://dx.doi.org/10.1016/j.cej.2009.10.055

[138] RODRÍGUEZ, N.; ALONSO, M.; ABANADES, C. J.: Experimental investigation of a cir-
       culating fluidized-bed reactor to capture CO2 with CaO. *AIChE Journal* 57 (5), p.
       1356–1366, 2011. DOI: http://dx.doi.org/10.1002/aic.12337

[139] ROMANO, M. C.: Modeling the carbonator of a Ca-looping process for CO2 capture from power plant flue gas. *Chemical Engineering Science* 69 (1), p. 257–269, 2012. DOI: http://dx.doi.org/10.1016/j.ces.2011.10.041

[140] ROMANO, M. C.: Ultra-high CO2 capture efficiency in CFB oxyfuel power plants by calcium looping process for CO2 recovery from purification units vent gas. *International Journal of Greenhouse Gas Control* 18, p. 57–67, 2013. DOI: http://dx.doi.org/1 0.1016/j.ijggc.2013.07.002

[141] ROMANO, M. C.; SPINELLI, M.; CAMPANARI, S.; CONSONNI, S.; CINTI, G.; MARCHI, M.; BORGARELLO, E.: The Calcium Looping Process for Low CO2 Emission Cement and Power. *Energy Procedia* 37, p. 7091–7099, 2013. DOI: http://dx.doi.org/10.1016/j.egy pro.2013.06.645

[142] ROMEO, L. M.; ABANADES, C. J.; ESCOSA, J. M.; PAÑO, J.; GIMÉNEZ, A.; SÁNCHEZ-BIEZMA, A.; BALLESTEROS, J. C.: Oxyfuel carbonation/calcination cycle for low cost CO2 capture in existing power plants. *Energy Conversion and Management* 49 (10), p. 2809–2814, 2008. DOI: http://dx.doi.org/10.1016/j.enconman.2008.03.022

[143] ROMEO, L. M.; CATALINA, D.; LISBONA, P.; LARA, Y.; MARTÍNEZ, A.: Reduction of greenhouse gas emissions by integration of cement plants, power plants, and CO2 capture systems. *Greenhouse Gases: Science and Technology* 1 (1), p. 72–82, 2011. DOI: http://dx.doi.org/10.1002/ghg3.5

[144] ROMEO, L. M.; LISBONA, P.; LARA, Y.; MARTÍNEZ, A.: Energy and exergy pertaining to solid looping cycles: Calcium and Chemical Looping Technology for Power Generation and Carbon Dioxide (CO2) Capture. Elsevier, 2015. ISBN: 9780857092434. DOI: http://dx.doi.org/10.1016/B978-0-85709-243-4.00002-1

[145] RYU, H.-J.; GRACE, J. R.; LIM, C. J.: Simultaneous CO2/SO2 Capture Characteristics of Three Limestones in a Fluidized-Bed Reactor. *Energy & Fuels* 20 (4), p. 1621–1628, 2006. DOI: http://dx.doi.org/10.1021/ef050277q

[146] SCALA, F.; CAMMAROTA, A.; CHIRONE, R.; SALATINO, P.: Comminution of limestone during batch fluidized-bed calcination and sulfation. *AIChE Journal* 43 (2), p. 363–373, 1997. DOI: http://dx.doi.org/10.1002/aic.690430210

[147] SCALA, F.; SALATINO, P.; BOEREFIJN, R.; GHADIRI, M.: Attrition of sorbents during fluidized bed calcination and sulphation. *Powder Technology* 107 (1-2), p. 153–167, 2000. DOI: http://dx.doi.org/10.1016/S0032-5910(99)00185-0

[148] SCEATS, M.: Oxide products formed from calcined carbonate powder for use as biocide, chemical detoxifier and catalyst support products, United States, Patent: WO/2015/100468, 10.12.2014

[149] SCEATS, M.; HORLEY, G.: System and method for calcination/carbonation cycle processing, Patent: WO/2007/045048, 23.10.2006

[150] SCHORCHT, F.; KOURTI, I.; SCALET, B. M.; ROUDIER, S.; DELGADO SANCHO, L.: Best available techniques (BAT) reference document for the production of cement, lime and magnesium oxide Industrial Emissions Directive 2010/75/EU (integrated pollution prevention and control). Publications Office, 2013. ISBN: 978-92-79-32944-9. DOI: http://dx.doi.org/10.2788/12850

[151] SEDGHKERDAR, M. H.; MAHINPEY, N.; ELLIS, N.: The effect of sawdust on the calcination and the intrinsic rate of the carbonation reaction using a thermogravimetric analyzer (TGA). *Fuel Processing Technology* 106, p. 533–538, 2013. DOI: http://dx.doi.org/10.1016/j.fuproc.2012.09.024

[152] SEDGHKERDAR, M. H.; MOSTAFAVI, E.; MAHINPEY, N.: Investigation of the Kinetics of Carbonation Reaction with Cao-Based Sorbents Using Experiments and Aspen Plus Simulation. *Chemical Engineering Communications* 202 (6), p. 746–755, 2015. DOI: http://dx.doi.org/10.1080/00986445.2013.871709

[153] SHAH, M.: Carbon dioxide (CO2) compression and purification technology for oxy-fuel combustion: Oxy-Fuel Combustion for Power Generation and Carbon Dioxide (CO2) Capture. Elsevier, 2011. ISBN: 9781845696719. DOI: http://dx.doi.org/10.1533/9780857090980.2.228

[154] SHAH, M.; DEGENSTEIN, N.; ZANFIR, M.; KUMAR, R.; BUGAYONG, J.; BURGERS, K.: Near zero emissions oxy-combustion CO2 purification technology. *Energy Procedia* 4, p. 988–995, 2011. DOI: http://dx.doi.org/10.1016/j.egypro.2011.01.146

[155] SHIMIZU, T.; HIRAMA, T.; HOSODA, H.; KITANO, K.; INAGAKI, M.; TEJIMA, K.: A Twin Fluid-Bed Reactor for Removal of CO2 from Combustion Processes. *Chemical Engineering Research and Design* 77 (1), p. 62–68, 1999. DOI: http://dx.doi.org/10.1205/0 26387699525882

[156] SILABAN, A.; HARRISON, D. P.: High temperature capture of carbon dioxide: characteristics of the reversible reaction between CaO(s) and CO2(g). *Chemical Engineering Communications* 137 (1), p. 177–190, 2007. DOI: http://dx.doi.org/10.1080/009864495 08936375

[157] SILABAN, A.; NARCIDA, M.; HARRISON, D. P.: Characteristics of the reversible reaction between CO2(g) and calcined dolomite. *Chemical Engineering Communications* 146 (1), p. 149–162, 2007. DOI: http://dx.doi.org/10.1080/00986449608936487

[158] SNOW, M. J. H.; LONGWELL, J. P.; SAROFIM, A. F.: Direct sulfation of calcium carbonate. *Industrial & Engineering Chemistry Research* 27 (2), p. 268–273, 1988. DOI: http://dx.doi.org/10.1021/ie00074a011

[159] SOSA ENDARA, J. A.: Comparative analysis of Calcium Looping CO2 capture from power and cement production. Master thesis, University of Stuttgart, 2017

[160] STARK, J.; WICHT, B.: Zement und Kalk Der Baustoff als Werkstoff. Birkhäuser Basel, 2000. ISBN: 978-3-7643-6216-4. DOI: http://dx.doi.org/10.1007/978-3-0348-8382-5

[161] STENDARDO, S.; FOSCOLO, P. U.: Carbon dioxide capture with dolomite: A model for gas–solid reaction within the grains of a particulate sorbent. *Chemical Engineering Science* 64 (10), p. 2343–2352, 2009. DOI: http://dx.doi.org/10.1016/j.ces.2009.02.009

[162] SUN, P.; GRACE, J. R.; LIM, C. J.; ANTHONY, E. J.: Removal of CO2 by Calcium-Based Sorbents in the Presence of SO2. *Energy & Fuels* 21 (1), p. 163–170, 2007. DOI: http://dx.doi.org/10.1021/ef060329r

[163] SUN, P.; GRACE, J. R.; LIM, C. J.; ANTHONY, E. J.: The effect of CaO sintering on cyclic CO2 capture in energy systems. *AIChE Journal* 53 (9), p. 2432–2442, 2007. DOI: http://dx.doi.org/10.1002/aic.11251

[164] SUN, P.; GRACE, J. R.; LIM, C. J.; ANTHONY, E. J.: A discrete-pore-size-distribution-based gas–solid model and its application to the CaO + CO2 reaction. *Chemical Engineering Science* 63 (1), p. 57–70, 2008. DOI: http://dx.doi.org/10.1016/j.ces.2007.08.054

[165] Sun, P.; Grace, J. R.; Lim, C. J.; Anthony, E. J.: Determination of intrinsic rate constants of the CaO–CO2 reaction. *Chemical Engineering Science* 63 (1), p. 47–56, 2008. DOI: http://dx.doi.org/10.1016/j.ces.2007.08.055

[166] Swift, W. M.; Panek, A. F.; Smith, G. W.; Vogel, G. J.; Jonke, A. A.: Decomposition of calcium sulfate: a review of the literature. United States, 1976. DOI: http://dx.doi .org/10.2172/7224692

[167] Szekely, J.; Evans, J. W.: A structural model for gas–solid reactions with a moving boundary. *Chemical Engineering Science* 25 (6), p. 1091–1107, 1970. DOI: http://dx.d oi.org/10.1016/0009-2509(70)85053-9

[168] Szekely, J.; Evans, J. W.; Sohn, H. Y.: The Elements of Gas–Solid Reaction Systems Involving Single Particles: Gas-solid Reactions. Elsevier, 1976. ISBN: 9780126808506. DOI: http://dx.doi.org/10.1016/B978-0-12-680850-6.50007-X

[169] Talib, Q.: Experimental investigation of a tail-end Calcium Looping CO2 capture process for cement application. Master thesis, University of Stuttgart, 2018

[170] Thormann, P.: Importance of separate grinding of limy clay and sand for clinker mineral formation during cement burning. *Tonind.-Ztg. Keram. Rund-schau* 92 (1), p. 7–11, 1968

[171] Valverde, J. M.: A model on the CaO multicyclic conversion in the Ca-looping process. *Chemical Engineering Journal* 228, p. 1195–1206, 2013. DOI: http://dx.doi.o rg/10.1016/j.cej.2013.05.023

[172] Verein Deutscher Zementwerke e.V.: Zement-Taschenbuch. Verlag Bau und Technik, 2002. ISBN: 3764004274

[173] Vincent, A.; Rennie, D.; Sceats, M.; Gill, M.; Thomsen, S.: Public LEILAC pre-FEED Summary Report. LEILAC Consortium, 2017. https://ec.europa.eu/research/pa rticipants/documents/downloadPublic?documentIds=080166e5ade8a49d&appId=P PGMS

[174] Voldsund, M.; Gardarsdottir, S.; Lena, E. de; Pérez-Calvo, J.-F.; Jamali, A.; Berstad, D.; Fu, C.; Romano, M.; Roussanaly, S.; Anantharaman, R.; Hoppe, H.; Sutter, D.; Mazzotti, M.; Gazzani, M.; Cinti, G.; Jordal, K.: Comparison of Technologies for CO2 Capture from Cement Production—Part 1: Technical Evaluation. *Energies* 12 (3), p. 559, 2019. DOI: http://dx.doi.org/10.3390/en12030559

[175] VOVELLE, C.: Pollutants from Combustion Formation and Impact on Atmospheric Chemistry. Springer Netherlands, 2000. ISBN: 978-94-011-4249-6. DOI: http://dx.doi .org/10.1007/978-94-011-4249-6

[176] WANG, J.; ANTHONY, E. J.: On the Decay Behavior of the CO2 Absorption Capacity of CaO-Based Sorbents. *Industrial & Engineering Chemistry Research* 44 (3), p. 627–629, 2005. DOI: http://dx.doi.org/10.1021/ie0493154

[177] WANG, Y.; LIN, S.; SUZUKI, Y.: Study of Limestone Calcination with CO2 Capture Decomposition Behavior in a CO2 Atmosphere. *Energy & Fuels* 21 (6), p. 3317–3321, 2007. DOI: http://dx.doi.org/10.1021/ef700318c

[178] WANG, Y.; LIN, S.; SUZUKI, Y.: Limestone Calcination with CO2 Capture (II): Decomposition in CO/Steam and CO2 /N2 Atmospheres. *Energy & Fuels* 22 (4), p. 2326–2331, 2008. DOI: http://dx.doi.org/10.1021/ef800039k

[179] WHITE, V.; ALLAM, R.; MILLER, E.: Purification of Oxyfuel-Derived CO2 for Sequestration or EOR. In: GALE, J.; BOLLAND, O.: 8th International Conference on Greenhouse Gas Control Technologies, 19-22 June, 2006, Trondheim, Norway. Curran, 2006. ISBN: 9781605603537

[180] WILCOX, J.: Carbon Capture. Springer US, 2012. ISBN: 978-1-4614-2215-0. DOI: http: //dx.doi.org/10.1007/978-1-4614-2215-0

[181] WILHELMSSON, B.; KOLLBERG, C.; LARSSON, J.; ERIKSSON, J.; ERIKSSON, M.: A feasibility study evaluating ways to reach sustainable cement production via the use of electricity. CemZero Consortium, . https://www.cementa.se/sites/default/files/ass ets/document/65/de/final_cemzero_2018_public_version_2.0.pdf.pdf (Access on 01.03.2021)

[182] WOLFRUM, H.: Demonstration des Calcium-Looping-Prozesses zur CO2-Abscheidung in der Zementindustrie. Diploma thesis, University of Stuttgart, 2016

[183] YLÄTALO, J.; PARKKINEN, J.; RITVANEN, J.; TYNJÄLÄ, T.; HYPPÄNEN, T.: Modeling of the oxy-combustion calciner in the post-combustion calcium looping process. *Fuel* 113, p. 770–779, 2013. DOI: http://dx.doi.org/10.1016/j.fuel.2012.11.041

[184] YLÄTALO, J.; RITVANEN, J.; ARIAS, B.; TYNJÄLÄ, T.; HYPPÄNEN, T.: 1-Dimensional modelling and simulation of the calcium looping process. *International Journal of Greenhouse Gas Control* 9, p. 130–135, 2012. DOI: http://dx.doi.org/10.1016/j.ijggc.2 012.03.008

[185] YU, F.-C.; FAN, L.-S.: Kinetic Study of High-Pressure Carbonation Reaction of Calcium-Based Sorbents in the Calcium Looping Process (CLP). *Industrial & Engineering Chemistry Research* 50 (20), p. 11528–11536, 2011. DOI: http://dx.doi.org/10. 1021/ie200914e

[186] ZHANG, Y.; GONG, X.; CHEN, X.; YIN, L.; ZHANG, J.; LIU, W.: Performance of synthetic CaO-based sorbent pellets for CO2 capture and kinetic analysis. *Fuel* 232, p. 205–214, 2018. DOI: http://dx.doi.org/10.1016/j.fuel.2018.05.143

[187] ZHOU, Z.; XU, P.; XIE, M.; CHENG, Z.; YUAN, W.: Modeling of the carbonation kinetics of a synthetic CaO-based sorbent. *Chemical Engineering Science* 95, p. 283–290, 2013. DOI: http://dx.doi.org/10.1016/j.ces.2013.03.047

Printed in the United States
by Baker & Taylor Publisher Services